POWER, INFLUENCE TACTICS, AND INFLUENCE PROCESSES IN VIRTUAL TEAMS

by

Marla Boughton

A dissertation submitted to the faculty of
The University of North Carolina at Charlotte
in partial fulfillment of the requirements
for the degree of Doctor of Philosophy in
Organizational Science

Charlotte

2011

Approved by:

_____
Dr. Anita L. Blanchard

_____
Dr. Lisa S. Walker

_____
Dr. Shawn D. Long

_____
Dr. Kelly L. Zellars

UMI Number: 3474310

All rights reserved

INFORMATION TO ALL USERS
The quality of this reproduction is dependent on the quality of the copy submitted.

In the unlikely event that the author did not send a complete manuscript
and there are missing pages, these will be noted. Also, if material had to be removed,
a note will indicate the deletion.

Dissertation Publishing

UMI 3474310
Copyright 2011 by ProQuest LLC.
All rights reserved. This edition of the work is protected against
unauthorized copying under Title 17, United States Code.

ProQuest LLC.
789 East Eisenhower Parkway
P.O. Box 1346
Ann Arbor, MI 48106 - 1346

© 2011
Marla Diane Boughton
ALL RIGHTS RESERVED

ABSTRACT

MARLA DIANE BOUGHTON. Power, influence tactics, and influence processes in virtual teams. (Under direction of DR. ANITA L. BLANCHARD)

Current studies of power, influence tactics, and influence processes in virtual teams assume that these constructs operate in a similar manner as they do in the face-to-face (FtF) environment. However, the virtual context differs from the FtF environment on a variety of dimensions, such as the availability of status cues. The differences between these contexts may alter how power and influence tactics are expressed in virtual teams. This study examines how power, influence tactics, and influence processes are manifested in virtual teams and which influence tactics are most successful in this context.

Twenty-three members of virtual teams were interviewed about their previous attempts to influence team members. The data were coded using a thematic approach. The main findings of the current study were: 1) There is a tendency to use more assertive influence tactics in virtual teams; 2) The success rate of influence tactics varies by the direction of the influence attempt, with lateral influence tactics having the lowest likelihood of success; 3) Specific status characteristics such as knowledge and skills are more relevant for members of virtual teams than diffuse status characteristics; and 4) There is both a relationship orientation and a task orientation in virtual teams.

I also present a model for the influence process in virtual teams. First, actors must use technology to get their targets' attention. Second, actors should build relationships through getting to know one another and the establishing trust, although this is not a requisite step. Third, actors must choose which influence tactic to use. While many choose to adapt traditional tactics to work in the virtual environment, new influence tactics (e.g., ambiguity reduction techniques) have emerged. Communication technology preferences affect which technologies an actor uses to build relationships and enact influence tactics. The status of the actor and target also affect which influence tactic(s) an actor uses.

Recommendations are offered for both low-status members of virtual teams as well as virtual team leaders. Members of virtual teams need to be more assertive in their influence attempts and also need to focus on building relationships with their team members in order to be successful influencers. Future research opportunities are also discussed. Given the growing prevalence of virtual teams, the results of this study are a valuable contribution to both practice and research.

DEDICATION

This dissertation is dedicated to my family, friends, and anyone who believed in me even when I doubted myself. I am especially grateful to John, who has been by my side throughout this entire process. Without you, I would not be where I am today.

ACKNOWLEDGEMENTS

First and foremost, I would like to thank my advisor, Dr. Anita Blanchard, who spent countless hours discussing ideas, proofing drafts, and providing feedback throughout the dissertation process. It is an understatement to say that this project would not have been possible without her support. Beyond the immeasurable academic contributions she has made to my research stream, she has also been my role model, my advocate, and a source of emotional support throughout this arduous process. In addition, I would like to thank the rest of my dissertation committee Dr. Lisa Walker, Dr. Shawn Long, and Dr. Kelly Zellars, for their insightful ideas and support. Each of them has brought a unique and valuable perspective to this project which has improved it beyond what I could have done on my own.

I would like to acknowledge the assistance of the Organizational Science program which provided me with funds to incentivize people to participate in the study. The program also provided me with the equipment necessary to transcribe the interviews that were collected. Finally, for their assistance during the data collection process, I would like to acknowledge the administrator of *Virtual Team Builders*, Ms. Claire Sookman, and the members of ISPI Charlotte.

TABLE OF CONTENTS

INTRODUCTION	1
Virtual Teams	2
Power	4
Influence Tactics	7
Leadership in Virtual Teams	8
Power and Influence Tactics in Virtual Teams	11
Expectation States Theory and Status Characteristics Theory	13
Status Characteristics and Status Cues	15
Status Characteristics and Influence Tactics	16
Summary	16
Status Characteristics Theory in a Virtual Context	17
Positive Effects of CMC on Status	19
Negative Effects of CMC on Status	20
Anonymity in Computer-mediated Environments	22
Status Characteristics and Status Cues in Computer-mediated Environments	24
The Current Study	29
METHOD	31
Participants	31
Materials	31
Questionnaire	31
Interview protocol	32

Procedure	32
Analysis Strategy	34
RESULTS	36
Influence Tactics in Virtual Teams	36
Traditional FtF Influence Tactics Also Present in Virtual Teams	37
Pressure	37
Legitimating Tactics	39
Rational Persuasion	41
Consultation	42
Inspirational Appeals	44
Exchange	45
Ingratiation	45
Coalition Tactics	46
Personal Appeals	46
Ambiguity Reduction Techniques	47
Sharing Information	48
Creating Accountability	48
Giving Examples	49
Summary	50
Directional Differences in the Success of Influence Tactics in Virtual Teams	51
Upward Influence Tactics	51
Lateral Influence Tactics	52
Downward Influence Tactics	53

Overall Analysis	53
Additional Strategies Used in the Influence Process	56
Strategic Use of Technology to Get Attention	56
Multiple Communications	56
Use the Target's Technology of Choice	58
Use of Instant Messaging to Check Availability	59
Summary	59
Influence Strategies in Virtual Teams	59
Building Relationships	60
Getting to know others & putting a face with a name	61
Establishing trust	62
Team building & building a sense of community	62
Documenting Communications	64
Summary	65
Status Characteristics Theory in Virtual Teams	65
Status Cues	66
Status Characteristics	68
Diffuse Status Characteristics	68
Specific Status Characteristics	73
Summary	80
How Status Affects the Influence Process in Virtual Teams	80
Summary	82
Additional Findings	83

Communication Technology Preferences	83
In-group – Out-group Effects	85
Culture	86
Time	87
Time zone differences	88
Pace	89
Response time	89
Deadlines	90
Multitasking	91
Summary	91
Chapter Summary	91
DISCUSSION	93
Tendency to Use Assertive Influence Tactics	94
Variation in Success Rate of Influence Tactics	97
Increased Relevance of Specific Status Characteristics	99
A Dual Orientation	102
Relationship Orientation	103
Task Orientation	104
Summary	105
The Influence Process in Virtual Teams	106
Practical Implications	110
Limitations and Future Research	111
Conclusion	113

REFERENCES	117
TABLES	126
FIGURES	140
APPENDIX	141

INTRODUCTION

Power and influence are a part of every social interaction (Kemper & Collins, 1990). This is also true in the context of virtual teams, which—since physical interactions are limited in these types of teams—are based almost entirely on social interactions. By not understanding the unique contributions of the virtual environment to the influence process and the choice of influence tactics (i.e., the methods virtual team members use to gain and maintain power), we are neglecting an essential component of virtual team interactions. As more and more organizations use virtual teams as a way to save money and bring together experts who are geographically dispersed (Connaughton & Shuffler, 2007), it becomes even more important to understand what influence tactics and influence processes members of virtual teams use and which are successful.

While researchers have begun to look at this question, they generally fail to acknowledge the possibility that the virtual context may provide team members with opportunities to use unique influence tactics and influence processes that are different from or in addition to those proven effective in face-to-face (FtF) teams or relationships (e.g., Elron & Vigoda-Gadot, 2006). For example, high-status members of virtual teams may ignore the contributions of low-status members by not responding to their technology-based communications (i.e., emails, faxes, voicemails, etc) (Metiu, 2006). While ignoring individuals is also possible in FtF teams, it is easier for members of virtual teams to do so because they have a reasonable excuse; it is more socially acceptable to blame technological problems for lapses in communication and collaboration than to admit to purposefully ignoring team members. As another example, an individual can write an email or complete an assignment during normal business hours, but schedule the email or file to be sent at 11pm. This gives the impression that the person has worked harder and longer than her teammates, which can result in the individual receiving additional resources (e.g., more of the team leader's time and approval, a raise, etc). These resources can provide the individual with an increase in power.

Members of virtual teams may be developing new ways of influencing one another that are not allowed in FtF interactions. It is important for virtual team leaders to understand how members of their virtual teams influence one another so that they can take steps to encourage lower status team members to participate in brainstorming and decision making. By creating more equal influence within the team, virtual team leaders are promoting better idea generation and more effective decisions (Bonito, DeCamp, & Ruppel, 2008; Janis & Mann, 1977). A primary purpose of this study is, therefore, to understand how power and influence tactics are manifested in virtual teams and why. A second purpose is to identify which influence tactics are most successful in virtual teams. That is, this study will identify which influence tactics are most effective in obtaining the desired results from a target. In addition, I will examine how status affects the influence process in virtual team interactions. In the following sections, I will define and review the relevant literature on virtual teams. I will then review theory and research on power, influence tactics, and status characteristics theory and also examine these constructs as they relate to virtual teams. Power and status characteristics theory are useful approaches because they explain how and why influence tactics are successfully used in virtual teams, and they suggest that status may play a role in the influence process.

Virtual Teams

Virtual teams are teams whose members are "mediated by time, distance, or technology" and whose members are interdependent, working together on a common task (Driskell, Radtke, Salas, 2003, p.297). Members of virtual teams communicate through various technologies including telephone, meeting room video conferencing, desktop video and audio conferencing, chat rooms for text interactions, file transfer, application sharing, and/or virtual reality options (Olson & Olson, 2000). These technologies vary as to how much they incorporate the media dimensions of copresence, visibility, audibility, contemporality, simultaneity, sequentiality, reviewability, and revisability (Clark & Brennan, 1991). These dimensions are important because they are associated with the richness of the media (i.e., the capabilities of the technology and how much information it conveys).

Virtual teams have become increasingly common in organizations (Connaughton Shuffler, 2007). As a result, research on virtual teams has become popular and covers a wide variety of topics including trust (Jarvenpaa, Knoll, & Leidner, 1998; Jarvenpaa & Leidner, 1999), effectiveness and performance (Kirkman, Rosen, Tesluk, & Gibson, 2004; Lurey & Raisinghani, 2001; Maznevski & Chudoba, 2000), conflict (Hinds & Bailey, 2003; Hinds & Mortensen, 2005), and communication and knowledge sharing (Constant, Sproull, & Kiesler, 1996; Cramton, 2001; Hinds & Kiesler, 1995). Leadership in virtual teams has also attracted research attention (Avolio, Kahai, & Dodge, 2000; Bell & Kozlowski, 2002; Kayworth & Leidner, 2000; Kayworth & Leidner, 2001-2002; Powell, Piccoli, & Ives, 2004; Zigurs, 2003). This area of research is useful for this study because leaders are given formal power within organizations. Indeed, leadership is a type of power (Zigurs, 2003). The research on virtual team leadership will be described in more detail when I review the literature on power and influence tactics in virtual teams.

Powell and her colleagues (2004) provide an in-depth review of the literature on virtual teams. Their review takes an input-process-output approach in which design, culture, technical expertise, and training are the inputs; relationship building, cohesion, and trust are the socio-emotional processes; communication, coordination, and task-technology-structure fit are the task processes; and performance and satisfaction are the outputs. Most relevant to the current study is the review of the socio-emotional processes present in virtual teams; relationship building enables individuals to establish personal power.

While Powell and her colleagues' (2004) review is informative, it does not adequately address power, influence tactics, and influence processes in virtual teams. In the following sections, I will describe theory and research on power, influence tactics, and influence processes in the FtF setting. I will then examine these constructs in the context of virtual teams and computer-medicated communication (CMC) theory.

Power

Perrow (1986) defines power as "the ability of persons or groups to extract for themselves valued outputs from a system in which the other persons or groups either seek the same outputs for themselves or would prefer to expend their effort toward other outputs" (p.259). The struggle for power is an inherent part of human interactions (Kemper & Collins, 1990). Powerful people command respect and are allowed to make decisions that affect others.

Power is a pervasive feature of organizational life. However, in this study I will limit my focus to the prescriptive nature of power as defined by French and Raven (1959). Within organizations, power allows individuals to gain resources and distribute rewards, punishments, and sanctions (French & Raven, 1959). Ultimately, power involves a person's ability to influence another person.

How do people get power? There are five bases of power: (1) reward power, the ability to determine how rewards will be distributed; (2) coercive power, the ability to punish others for not complying with a request; (3) legitimate power, the formal authority associated with organizational position; (4) referent power, the ability to influence others based on a person's desirable traits such as attractive personalities or appearance; and (5) expert power, the ability to influence others based on a person's expertise, skill, or knowledge (French & Raven, 1959; Vecchio, 1997). Legitimate, reward, and coercive power bases are more formal and are based on an individual's organizational position (Bass, 1960; Vecchio, 1997). I will refer collectively to these more formal bases of power as *formal power*. Referent and expert bases of power are more informal and stem from an individual's personal characteristics (Vecchio, 1997). I will refer collectively to the informal bases of power as *personal power* (Bass, 1960).

Recently, Baldwin, Kiviniemi, and Snyder (2009) found that an informational advantage (i.e., expert power)—a subtler, more implicit way of conferring power than by position (a type of formal power)—led participants to have stronger feelings of power compared to those without expectations of an informational advantage. This study identified informational advantage as another source of power. Informational advantage is most closely related to personal power, underscoring the important role personal power plays in interpersonal interactions.

In FtF organizations and the formal groups that exist within them, power is determined primarily by structure; the same power is given to whoever occupies the position (Weber, 1946). Personal power is also present in organizations, albeit to a lesser extent (Peiro & Melia, 2003). Members of organizations—those with and without formal power—can use their personal power to influence other organization members.

Also prevalent are informal groups in which there is no formal power but in which power, nonetheless, emerges. As there is no formal structure, members of informal groups use personal power based on their characteristics and behaviors (e.g., sensitivity to others or flexibility as conditions change) (Pfeffer, 1992). These characteristics can include expertise as well as desirable traits or resources that inspire admiration (e.g., likeability or charisma) (Bass, 1960; French & Raven, 1959).

Anderson, Spataro, & Flynn (2008) examined power in FtF organizations. Using survey research, they found that the attainment of power is dependent upon a person's fit with the organization, specifically in regards to their personality characteristics. They found that extraverts are more powerful in team-oriented organizations, whereas conscientious employees are more powerful when they work alone.

Expanding upon Anderson and colleagues' (2008) study, Anderson and Kilduff (2009) examined how dominant individuals gain influence in groups. In order to do this, the researchers brought together undergraduate students in groups of four to work on a task. Upon the completion of the task, participants rated their team members on influence, competence, and personality dimensions. Although dominance itself did not provide individuals with higher levels of influence, the results showed that dominant people were perceived by their peers as more socially competent, and were thus given more influence. Together, these two studies show the importance of personal power in organizations.

Although theories of power contribute a great deal to our understanding of the social interactions within organizations and teams, they do not describe how people actively seek to control others. In the following section, I will describe how people translate their power into action. In this section, I will also describe how individuals without formal power can, nonetheless, influence other people.

Influence Tactics

Influence tactics are the methods people use to translate power into action. One example of an influence tactic would be to use your expertise to lay out facts and statistics in order to persuade someone to use your solution or allow you to make a decision. Another example of an influence tactic would be to force someone to do something for you by threatening them with a punishment. In sum, influence tactics are behaviors that allow individuals to exert power, while power is an ability that arises from both organizational (e.g., position) and personal (e.g., expertise) sources.

Power and influence tactics are closely related constructs; they both enable individuals to influence others in an organizational setting (Anderson et al, 2008). Power and influence are both defined by changing the behavior of others; however, they differ in that power enables this control to be enacted more easily (Vecchio, 1997). To put it another way, influence is subtler and less reliable than power.

Influence tactics are used in formal groups and organizations as well as in informal groups and by people with and without formal power. Through influence tactics, power can be asserted in multiple directions (i.e., upward, downward, and lateral) (Yukl, Chavez, & Seifert, 2005; Yukl & Falbe, 1990); that is, influence tactics can help individuals have power over others. Although both the formal and personal bases of power are relatively stable (Bass, 1960), influence tactics enable *all* individuals, regardless of whether they have formal power, to influence others.

In their original work which built upon the Kipnis Schmidt, & Wilkinson (1980) study, Yukl and Falbe (1990) identified eight influence tactics: pressure tactics, upward appeals, exchange tactics, coalition tactics, ingratiating tactics, rational persuasion, inspirational appeals, and consultation tactics. In 2005, Yukl and his colleagues identified collaboration and apprising (i.e., explaining how the target person will benefit by complying) as two additional influence tactics. The frequency with which tactics are used varies with the direction of influence (i.e., upward, lateral, or downward) (Yukl & Falbe, 1990). The choice of influence tactic also depends on the situation or context. For example, it is not always appropriate or effective to use pressure tactics and ingratiation is more effective moving downward than upward (Yukl & Tracey, 1992). However, subtler influence tactics (e.g., consultation and rational persuasion) tend to be more successful, regardless of the direction of the influence tactic (Yukl & Falbe, 1990).

Leadership in Virtual Teams

How are power and influence tactics in virtual teams different than power and influence tactics in FtF teams? One place to begin answering this question is by looking at the leadership literature. Leaders have formal power. Leaders influence their followers.

Therefore, research on virtual team leaders can provide a starting point for examining peer-to-peer power and influence tactics in virtual teams. Theoretical works of virtual team leadership cover a wide variety of topics, including the differences between virtual team leadership and FtF leadership, the leadership traits relevant to virtual team leadership, the challenges of virtual team leadership, and how to overcome these challenges (Avolio et al, 2000; Bell & Kozlowski, 2002; Cascio, & Shurygailo, 2003; Zigurs, 2003). However, the work on virtual team leadership described below is most important for an understanding of power and influence tactics.

Zhang and Fjermestad (2006) theorize that—due to increases in organizational and cultural distance inherent in many virtual teams—leaders of virtual teams often lack sufficient legitimate power. For example, a lack of legitimate power (i.e., power based on position) can arise if members of the team are higher in the organizational hierarchy than the team leader. In addition, virtual team leaders may not have the ability to give rewards (i.e., reward power) if the organization does not give them resources for rewards (Kayworth & Leidner, 2000). Similarly, virtual team leaders may lack coercive power; that is, they may be unable to punish team members that come from other departments or organizations. While this lack of legitimate power could also occur in FtF teams, Zhang and Fjermestad (2006) assert that the effects will be exacerbated in the virtual context. Zhang and Fjermestad (2006) propose that when their formal power is limited, virtual team leaders will need to use personal power.

Research has supported Zhang and Fjermestad's (2006) assertion that personal relationships are an important part of virtual team leadership. Pauleen (2003) found that virtual team leaders consider building relationships with virtual team members to be essential to the success of the virtual team. In fact, most virtual team leaders felt that establishing these relationships was necessary before work on the task could even be commenced. Thus, virtual team leaders prefer not to rely entirely on their formal power and instead choose to use their personal characteristics to form personal relationships.

Yoo and Alavi (2004) examined emergent leaders in virtual teams and identified one way these leaders can influence their followers that has not been identified in the FtF literature. Using qualitative research, they found that virtual team leaders can influence team members by patterning temporal structures (e.g., scheduling conference calls). By patterning the temporal structure of the team, the emergent leader controlled when team members worked on tasks, when their assignments were due, and when the team would communicate with one another as a group, thus controlling the flow of information to a certain extent. The results of this study suggest that technology may provide unique ways to influence virtual team members.

In summary, these studies on virtual team leadership suggest that personal power may be more effective than formal power in virtual teams and that the structural differences inherent in virtual teams may be the reason why the effectiveness and efficiency of formal power are decreased in the virtual context. They also provide some evidence that the characteristics of virtual teams (e.g., distance and technology) may create opportunities for virtual team members to use influence tactics unique to the virtual environment.

Although power and influence tactics are an inherent part of virtual team leadership due to leaders' formal power and their influence over followers (Avolio et al, 2000), the use of influence is not confined to virtual team leaders. This study will contribute to the literature by expanding the study of power and influence in virtual teams from the downward influence of virtual team leaders to the peer-to-peer influence of virtual team members.

Power and Influence Tactics in Virtual Teams

Theory and research indicate that power and influence processes do not work in exactly the same way in virtual teams as they do in FtF teams. Avolio and Kahai (2003) point out that CMC provides everyone with the power to reach out and touch everyone (e.g., through email). Geographically distributed team members can also easily withhold information from one another (Rosen, Furst, & Blackburn, 2007). These circumstances suggest that CMC provides at least a small amount of power to all virtual team members. The ability to easily contact everyone and withhold information provides virtual team members with a power that is unique to the virtual environment.

In what other ways do power, influence tactics, and influence processes differ in the virtual environment? Formal power may be constrained in the virtual environment due to increases in organizational and cultural distance (Zhang & Fjermestad, 2006). In addition, the use of formal power may lead to decreases in trust and cohesion in virtual teams. Zhang and Fjermestad (2006) suggest that, for these reasons, personal power may be more effective than formal power in the context of virtual teams.

Reliance on personal power, as opposed to formal power or authority, suggests that members of virtual teams will use influence tactics to exert power. The limited empirical work that has been conducted on influence in virtual teams supports this assertion. For example, group decision support systems are interactive networks of computers used to facilitate decision making and solution generation for unstructured problems in virtual teams (Sosik, Avolio, & Kahai, 1997). Studies of group decision support systems (Zigurs, Poole, & DeSanctis, 1988) and group support systems (Rains, 2005) suggest that virtual team members may try to influence each other more than in FtF teams. This is because decision support systems and group support systems create greater opportunities for peer member influence through its minimization of barriers to communication (Rains, 2005).

Recently, Elron and Vigoda-Gadot (2006) interviewed members of global virtual teams in order to examine how technology mediation affects influence and political processes. The researchers found that limited familiarity with team members was associated with the use of fewer influence tactics; additionally, the influence tactics that were used were softer (i.e., less obvious and forceful). They also found that membership in the virtual team was less central to participants' organizational identity and performance than membership in collocated teams, which made influencing members of the virtual team less important than influencing members of their FtF teams. Elron and Vigoda-Gadot concluded that FtF influence tactics are present in virtual teams, but they are less obvious than they are in FtF interactions.

Elron and Vigoda-Gadot (2006) contend that influence attempts are attenuated (i.e., softer) in virtual teams; however, social information processing theory suggests that power relations may develop over time (c.f., Walther, 1995). In support of this proposition, Saunders, Robey, and Vaverek (1994) found that status differences between physicians, hospital administrators, and nurses using computer conferencing became more established over time. If Elron and Vigoda-Gadot had studied members of well-established virtual teams, influence attempts may have been harder and more prevalent.

Once a team history is established and members are more comfortable communicating with one another, power plays and the use of influence tactics may become more common. Indeed, given the ambiguous nature of authority in the virtual environment (cf. Zhang & Fjermestad, 2006), influence tactics may occur more frequently than in FtF teams (i.e., ambiguity may make influential behavior more acceptable or less prone to sanctioning).

Previous research on power and influence tactics in virtual teams underestimates the importance of power in social interactions. Some individuals have a need for power that motivates their behavior across settings (deCharms & Muir, 1978). Therefore, it is unwise to assume that power is less important in the virtual environment than it is in FtF interactions. This assumption is especially ill-advised in the case of organizations, in which departments and individuals are constantly striving to gain power and resources. As virtual teams are nested within organizations, it makes intuitive sense that power will be important in this context as well.

One theory that explains power in social interactions is status characteristics theory. In the following section I will describe status characteristics theory and how it relates to power and influence. Then I will apply status characteristics theory to the virtual context, providing empirical examples when they are available. The concept of power, along with status characteristics theory, provides the conceptual framework of this study.

Expectation States Theory and Status Characteristics Theory

Expectation states theory provides an explanation as to how power emerges and is maintained in groups and teams (Wagner & Berger, 1993, 2002). Expectation states theory is a theoretical research program that encompasses a variety of interrelated theories that explain how social interactions are shaped by the expectations that one actor has for another actor (Wagner & Berger, 2002). These expectations reflect an actor's anticipation of how people will act in a given situation. For example, most people have higher performance expectations for college graduates than they do for high school dropouts, and this will affect how they treat individuals (i.e., how much responsibility they will give them, whether they will solicit their opinions, etc.).

Status characteristics theory—a part of the expectations states theoretical research program—addresses how initial status differences result in expectations for the immediate situation (Wagner & Berger, 1993). Previous research has shown a mutually reinforcing relationship between power and status in which power leads to status and status leads to power (Walker et al., 2000; Willer, Lovaglia, & Markovsky, 1997). The relationship between status and power is made clearer when one considers that, compared to those with low status, high-status individuals are given more opportunities to perform, are more likely to solve problems, are evaluated positively by group members, and are less likely to be manipulated by others and change their opinions (Berger, Ridgeway, Fisek, & Norman, 1998).

Status characteristics theory is relevant to the current research because virtual teams can be composed of individuals from a wide variety of backgrounds given their geographic dispersion (Hertel, Geister, & Kondradt, 2005). Even if virtual team members are collocated, it is likely that they differ from one another on at least one status dimension. Because of their diverse membership, virtual teams are more likely to be composed of individuals who display initial status differences.

Status characteristics and status cues are important components of status characteristics theory. A status characteristic is a socially established attribute on which people are differentially evaluated (Berger, Cohen, & Zelditch, 1972; Wagner & Berger, 2002). Status cues, on the other hand, are verbal and nonverbal indicators of performance capacities. Status characteristics theory argues that expectations are shaped by both status characteristics and status cues, which highlight status inequalities. The next sections will further define status characteristics and status cues and describe how they are related to influence tactics.

Status Characteristics and Status Cues

Status characteristics can be either diffuse or specific (Berger et al., 1972). Diffuse status characteristics are generalized assumptions about a specific population (Berger et al., 1972). Race, gender, ethnicity, and attractiveness are examples of diffuse status characteristics. Initial status differences are created based upon diffuse status characteristics that are stable and pervasive (Wagner & Berger, 2002).

Specific status characteristics (e.g., math ability or occupation) also exist (Berger et al., 1972). They are characteristics used to differentially evaluate people on their ability to succeed. Both diffuse and specific status characteristics determine which group members participate, have influence, and have prestige (Berger et al., 1972).

Expectations are not formed solely based on status characteristics; status cues also affect expectations. Status cues are the verbal and nonverbal cues upon which individuals base attributions of status and performance (Wagner & Berger, 2002). Examples of status cues include social cues such as patterns of speech, style of dress, and nonverbal cues such as posture and gestures (Wagner & Berger, 1993).

Status Characteristics and Influence Tactics

Status characteristics and influence tactics both affect an individual's ability to obtain and maintain power. How, then, are status and influence tactics related to one another? One possibility is that influence tactics are constrained by status characteristics. In other words, the success of an influence tactic may vary based upon the status of the person employing it. High-status individuals will, in general, be more successful at implementing influence tactics, and, as a result, will have more power than low-status individuals. Another possibility, suggested by Vecchio (1997), is that status is a type of influence tactic and that those who have higher status—or simply appear high in status—exert greater influence.

One purpose of this study is to examine how status affects influence tactics and influence processes in virtual teams. As an example, high-status individuals will be more likely to have formal power in organizations (Walker et al., 2000; Willer et al, 1997). This power enables high-status individuals to use the influence tactics of legitimacy (i.e., reliance on organizational position) and pressure (i.e., threats, demands, or warnings) (French & Raven, 1959; Raven, 1992; Yukl et al, 2005; Yukl & Falbe, 1990). In addition, a high-status individual has more success using influence tactics such as inspirational appeals and exchange (Yukl & Falbe, 1990; Yukl & Tracey, 1992).

Summary

In this section, I described expectation states theory and status characteristics theory. I also explained the difference between status and influence and how they relate to power. In the following section, I will review research and theory on status characteristics theory as it pertains to virtual teams.

Status Characteristics Theory in a Virtual Context

One question this paper seeks to answer is: How does the virtual environment change the influence processes within teams? Electronic media possess new capabilities which make electronic communication qualitatively different from traditional communication (Culnan & Markus, 1987; Driskell et al, 2003; Hambley, O'Neill, & Kline, 2007; Markus, 1994; Tidwell & Walther, 2002). Indeed, CMC has the ability to increase the amount of control a high-status person has over a low-status person by allowing the high-status person to observe others (e.g., monitoring emails or message content) (Spears & Lea, 1994). In addition, Driskell and his colleagues (2003) propose that status characteristics may have a greater impact in the virtual environment because cues are restricted, thus highlighting certain status characteristics while dampening others. In other words, in the virtual environment, individuals base their expectations of others on fewer, more prominent status characteristics than they do in the FtF environment. Therefore, it is reasonable to expect that electronic communication has the capacity to alter how individuals influence one another. This study will contribute to the literature by identifying how electronic media change how power is expressed by team members and what differences between FtF and virtual influence processes exist. In order to accomplish this goal, the current study will apply status characteristics theory to virtual teams.

Much of the previous theoretical work on virtual teams has been built around identity (i.e., SIP: Walther, 1992, 1996; SIDE: Reicher, Spears, & Postmes, 1995; Spears & Lea, 1994). Status characteristics theory—the theoretical framework of this study—provides a unique contribution to research on power and influence in virtual teams. It goes beyond explaining how team members perceive and categorize themselves—what identity theories do—and describes how team members fit in the group and how they interact with one another. Applying status characteristics theory to virtual teams allows us to advance the research on virtual teams by providing a framework for understanding power and influence tactics in teams as processes rather than as static constructs.

Within the past 25 years, researchers have examined status in the computer-mediated environment. Very few of these studies have used status characteristics theory as the basis for their research, and those that have used it found conflicting results (e.g., Dubrovsky, Kiesler, & Sethna, 1991; Hollingshead, 1996; Weisband, Schneider, & Connolly, 1995). This study will use status characteristics theory to clarify how power and influence processes function in a distributed setting.

One reason scholars are so interested in status in a computer-mediated or distributed environment is the dearth of status characteristics and status cues present in this context. According to status characteristics theory, expectations of group members are formed by status characteristics and status cues (Berger et al, 1972; Wagner & Berger, 2002); this invites the questions: how are expectations in virtual or distributed teams shaped? How (i.e., on what basis) is power distributed among members of these teams? Some researchers have argued that anonymity will lead to an equalization of status (Dubrovsky et al, 1991), while others have argued that status differences persist and may even be accentuated (Weisband et al, 1995). In the following section, I will review the literature on status in CMC, discuss the effects of anonymity on status in computer-mediated environments, and describe the various status characteristics and status cues that are available in virtual teams.

Positive Effects of CMC on Status

Two different schools of thought have emerged regarding how status functions in a computer-mediated environment. The first has been called the *equalization phenomenon* or the *benevolence hypothesis*. This concept was introduced 25 years ago when research on CMC was in its infancy. Kiesler, Siegel, and McGuire (1984) proposed that high-status people will have less influence and there will be more equal participation when communication is computer-mediated; this proposition is due to the fact that status is communicated neither contextually nor dynamically (i.e., through nonverbal behaviors).

The equalization phenomenon has its roots in the cues-filtered-out perspective. Theories from this perspective, such as social presence theory (Lind, 1999; Majchrzak, Rice, King, Malhotra, & Ba, 2000; Pauleen, 2003-2004; Walther & Burgoon, 1992; Warkentin & Beranek, 1999) and media richness theory (Daft & Lengel, 1986), claim that CMC is inherently impersonal due to a lack of nonverbal cues and that personal, intimate relationships cannot develop using CMC. Social presence theory and media richness theory argue that status differences are decreased in the virtual environment and formal power will not work as well as personal power in this context. However, these theories have been criticized as being flawed because CMC relationships can, in fact, be interpersonal and even hyperpersonal (see Walther, 1996).

Based in part upon Kiesler and her colleagues' (1984) assertion, the equalization phenomenon argues that anonymity is beneficial in CMC; like the cues-filtered-out approaches, the equalization phenomenon argues that the effects of status are attenuated or eliminated during computer interaction, allowing for more equal participation among group members (Dubrovsky et al, 1991). One explanation for this effect on status is the reduction in social-context cues that occurs with CMC. People feel less anxiety about being evaluated by higher status group members when status cues are not present (i.e., reduced evaluation anxiety), or they may even forget that there is another person receiving their typed messages (i.e., increased social inattention).

Empirical work has also been conducted which supports the equalization phenomenon. Sproull and Kiesler (1986) found evidence of status equalization in their study of electronic mail. There were no differences in message attributes (i.e., closing, positive affect, politeness, and energy) between messages sent by subordinates and those sent by superiors. In addition, participants preferred electronic mail over FtF interaction for upward communication. Using experimental methods, Siegel, Dubrovsky, Kiesler, & McGuire (1986) found that computer-mediated groups had more equal participation than groups that communicated FtF and asserted that this was evidence of social equalization.

Negative Effects of CMC on Status

The second school of thought on status in virtual teams consists of perspectives that argue that anonymity can have a negative impact on CMC. I have identified three different perspectives in the previous literature that support this point of view. The first is based on Foucault's metaphor of the *panopticon*. The panopticon perspective acknowledges that CMC has the power to both attenuate and accentuate status differences (Spears & Lea, 1994). Anonymity may lessen self-consciousness, feelings of responsibility, and regard for team members, leading to more equal participation. However, CMC also has the ability to increase the amount of control a high-status person has over a low-status person by enabling hierarchical observation.

The second perspective is known as *status persistence*. This perspective argues that, rather than eliminating status differences, CMC causes these differences to persist (Hollingshead, 1996). Several empirical studies have supported this perspective. For example, in their research, Silver, Cohen, and Crutchfield (1994) examined the effects of status differentiation on idea generation in computer-mediated groups. They found that high-status group members sent more words than low-status group members. This finding still held even though deindividuation (i.e., the attenuation of one's personal identity) and the reduction of status differences were observed. Thus, status affects the interactions of virtual team members.

In further support of this perspective, Saunders, Robey, and Valerek (1994) studied medical professionals engaged in computer conferencing and found that physicians and hospital administrators were given higher status than nurses. They also found that these status differences became more established over time. In an experiment comparing FtF groups with those that interact via CMC, Weisband and her colleagues (1995) found that status differences—as indicated by participation and influence—persisted in both FtF and computer-mediated groups. Pena, Walther, and Hancock (2007) argued that dominance perceptions (i.e., perceptions that others can elicit compliance or submission) vary with the level of available social information (e.g., status characteristics). In support of status persistence, they found that—rather than being eliminated—dominance perceptions (based on status) persisted and were more extreme in distributed than collocated groups.

Tan, Swee, Lim, Detenber, and Alsagoff (2008) found that status cues did not affect participation or perceptions of informativeness, persuasiveness, or source credibility (based upon expertise-related authority and character). While these findings lend credence to the equalization phenomenon, this study did find that status cues were affected by language-related authority and that language and expertise interacted to affect perceptions of the source's character. Thus, this study provides evidence that people are aware of subtle status cues and characteristics such as grammar and syntax in addition to showing that these status cues persist in the virtual environment.

The third perspective is the *discounting hypothesis*. The discounting hypothesis argues that, rather than leading to more equal participation by team members, anonymity undermines the spirit of technologies such as group decision support systems. This misrepresentation of the technology's spirit leads to unintended uses and outcomes (Rains, 2007). For example, without status cues, group members cannot judge a source's credibility, causing trust and consensus to decrease. As with the equalization phenomenon, support has been found for the discounting hypothesis. In their experimental research, McLeod and Liker (1992) found that electronic meeting systems did not affect participation equality; instead, they decreased task focus. Electronic meeting systems also led to lower performance on a complex generative task. Rains (2007) found that when perceptions of anonymity were controlled for, participants reported the anonymous confederate as less trustworthy, less persuasive and as having less goodwill toward the group. There was also a negative relationship to decision shifts and perceptions of the confederate's competence, regardless of the confederate's argument quality.

Anonymity in Computer-mediated Environments

All of the above perspectives are based upon an assumption of anonymity in computer-mediated settings. However, past research has shown that—even if present during initial interactions—anonymity dissipates over time (Walther, 1995). One theory that underlies this past research is social information processing theory (SIP). According to SIP, relationships among members of computer-mediated groups can reach the same level of intimacy as FtF relationships, regardless of the availability of nonverbal cues and identity cues.

SIP was developed as a response to the cues-filtered-out perspectives (Chidambaram, 1996; Chidambaram & Bostrom, 1993; Walther, 1995, 1996; Walther & Burgoon, 1992; Warkentin & Beranek, 1999) and argues that, rather than being impersonal, CMC is interpersonal (Walther, 1996). CMC enables social relationships to develop; however, as a result of the medium, relationships develop more slowly due to a difference in the rate of social information exchange. SIP contributes to the study of power, influence tactics, and influence processes in virtual teams because it shows that status and identity cues emerge in the virtual environment and interpersonal relationships are capable of forming in this context. Interpersonal relationships help individuals establish personal power.

Hyperpersonal theory (Walther, 1996) is an extension of SIP. This theory claims that CMC can lead to more intimate relationships than FtF communication because users present an optimized self and interpret others in an idealized manner (Walther, 1996). Thus, SIP and hyperpersonal theory argue against the cues-filtered-out approach.

Together SIP and hyperpersonal communication demonstrate that regardless of perceptions of anonymity in the virtual environment, identity cues—and thus status characteristics and status cues—are available in computer-mediated groups. Additionally, members of virtual teams communicate using a variety of channels of communication (i.e., type of media technology) (Driskell et al, 2003). These communication channels can include text, audio, video, or a combination. The type of technology used is important because, depending on the channel of communication used (e.g., text message, email, phone, SKYPE, video conference, etc.), status characteristics may or may not be available to members of virtual teams (See Table 1 and Table 2). In addition to the cues provided by the technologies used, virtual teams do not operate in a vacuum. Coworkers can share impressions about each other through their own FtF and social networks. In the following section, I will discuss the availability of various status characteristics and status cues in computer-mediated environments.

Status Characteristics and Status Cues in Computer-mediated Environments

Much of the previous research on status in virtual teams has studied them in the context of group decision support systems. These studies of group decision support systems assume that complete anonymity is a feature of the technology because contributors are not directly identified (Sosik et al, 1997). Under conditions of complete anonymity, diffuse status characteristics (e.g., gender and race) as well as specific status characteristics (e.g., expertise) are theoretically unavailable. Despite this purported anonymity, participants in group decision support systems can often identify individuals based on the evaluative tone of their comments and the amount of prior communication received from other group members (Hayne, Pollard, & Rice, 2003). Even if individuals inaccurately identify a fellow team member, they will still make attributions about team members' identities based on their assumption, and this, in turn, will affect their interactions (e.g., the decision making process) with the "identified" team member. Thus, status can prevail in conditions of "complete anonymity." The fact that status conditions are present even when anonymity is assumed further supports the importance of status characteristics theory in research on virtual teams.

In addition to the unintentional identifiers available in group decision support systems and other technologies, diffuse and specific status characteristics can be purposefully made available in virtual teams. Emails often contain signature files that divulge diffuse status characteristics. For example, a name in the signature file can inform the receiver of the sender's gender and, perhaps, age and ethnicity (or at least people will make attributions about age and ethnicity whether they are correct or not). Organizational position and education (e.g., executive vice president, administrative assistant, PhD, MD, etc) are often evident in signature files as well.

It cannot be assumed that all individuals who include information regarding their education in their signature files are high-status. While individuals who include their educational achievements in their signature file typically have advanced degrees (and thus higher status), those who include their organizational positions are not necessarily high-status individuals. For example, customer service representatives may include their titles in email communications with customers so that customers know with whom they are communicating.

Recently, employees in some organizations (e.g., Northwestern Mutual) have begun to include professional photographs in the signature file of their email. Photographs, in particular, contain diffuse status characteristics (e.g., gender, race, ethnicity, age, attractiveness) and nonverbal status cues (e.g., posture or style of dress), which are known to affect group interactions. Research has found that photographs affect computer-mediated interactions. For example, Walther, Slovacek, and Tidwell (2001) found that seeing one's communication partner was beneficial for new, unacquainted team members (i.e., it promoted affection and social attraction), but visual cues dampened interpersonal attraction in long-term computer-mediated groups most likely because they challenged participants' idealized virtual perceptions of their team members.

Channels that allow for verbal communication (i.e., telephone, SKYPE, conference calls, videoconference, etc.) can also convey the status characteristics of gender, race, or ethnicity. Tone of voice and accent are also available as status cues in the virtual environment. These cues convey status because people base their assumptions of others' performance and behavior on them (e.g., people who speak slowly are not as intelligent).

Nonverbal status cues (e.g., style of dress, posture, and gestures) (Wagner & Berger, 1993) also vary in their availability in the virtual setting. Expressive cues (e.g., posture)—from which people infer status—may become attenuated in a computer-mediated environment. However, indicative cues (e.g., members' stating their organizational positions) are direct labels of a person's status and will still be available for members of virtual teams. Task cues (e.g., fluency of speech or typing speed) provide information about a person's competency on the task at hand and may also be available in a virtual setting.

Theoretical work has explored the role of status in computer-mediated environments. Driskell and his colleagues (2003) developed an input-process-output model of the effects of CMC (input) on team performance (output). In this model, status is a proposed mediator between the input and the output. The researchers proposed three mechanisms through which technological mediation may impact status processes: 1) CMC may *block* the transmission of status characteristics and status cues; 2) the effects of status characteristics and status cues may be *dampened* in virtual environments; and 3) status expectations may not be *translated* into behavior due to weakened norms. This theoretical framework has its roots in the equalization phenomenon and the cues-filtered-out perspective.

Driskell and his colleagues (2003) also propose that the type of computer-mediated environment moderates the relationship between CMC and status processes (i.e., richer types of communication transmit more status characteristics and status cues). In support of the current study, the researchers propose that status characteristics may have a greater impact in virtual environments—which can highlight certain status characteristics while making others less salient—than in FtF environments.

In a recent ethnographic study, Metiu (2006) examined status dynamics in virtual groups. Status differences in this study were defined by geographic location rather than by individual differences. Team members in the United States were afforded high status and team members in India were afforded low status. This study provides an example of the us-versus-them mentality that can emerge in geographically distributed teams, thus causing status differences.

Metiu (2006) identified several "closure strategies" that high-status workers used to assert and maintain their status within the virtual team. Closure occurs when high-status team members monopolize opportunities and resources at the expense of low-status team members (Metiu, 2006). Closure strategies used at the group-level can be likened to influence tactics used at the individual level; both are used by people to maintain or increase status and power.

Examples of closure identified in this study include avoidance strategies available through CMC such as a lack of interaction with members of the low-status group and the use of the geographic boundaries (e.g., sending incomplete documents to India). More active closure strategies that were identified include: nonuse of work performed by the low-status group, criticism of the work performed by the low-status group, and the transfer of code ownership to the high-status group. While the closure strategies identified by Metiu may be available FtF, they are more easily enacted in virtual teams due to geographic separation and status differences (Metiu, 2006). In Metiu's (2006) study, the closure strategies served as a way to enhance the status of the high-status group and degrade the low-status group.

Metiu (2006) also found that team members in the United States manipulated technology as well as geographic boundaries in order to assert and maintain their status online. Metiu's (2006) findings support previous work that argues that electronic media possess new capabilities which makes electronic communication qualitatively different from traditional communication (Culnan & Markus, 1987; Driskell et al, 2003; Markus, 1994; Spears & Lea, 1994; Tidwell & Walther, 2002).

While Metiu's (2006) study advanced our understanding of status in virtual teams, it did not specifically look at power and influence tactics in this setting. This study will examine these constructs through the conceptual lens of status characteristics theory. In the following section I will describe the current study in more detail.

The Current Study

The first objective of this paper is to identify the influence tactics used in virtual teams. As a result, our methods will be adapted from those of researchers who identified influence tactics used in FtF interactions (c.f., Kipnis et al., 1980).

The fact that electronic media channels have more capabilities than FtF channels (Culnan & Markus, 1987) suggests that more (or simply different) influence tactics may be available to individuals who interact online as compared to those who interact FtF. Therefore, a primary objective of this study is to identify influence tactics unique to the virtual environment.

> RQ1a: What influence tactics are available to individuals who interact in virtual teams?
>
> RQ1b: How are these similar or different to those available to individuals who interact FtF?

In order to answer this question, this study will extend previous research that identifies influence tactics in FtF interactions to the virtual setting (cf., Kipnis et al, 1980; Yukl & Falbe, 1990).

The second objective of this study is to explain how the influence process varies among people of different statuses in a computer-mediated environment and how the computer-mediated influence process differs from the FtF influence process. For example, the influence process an individual uses may depend on whether they are high or low status in a given situation. This study will explore gender, ethnicity, age, tenure in the virtual team (e.g., new team member vs. old team member), and expertise as status characteristics that may affect how people influence one another in virtual teams.

RQ2a: How does status affect the influence process in virtual teams?

RQ2b: How do low-status individuals successfully exert power over other members of their virtual teams?

Ultimately, the goal of this study is to create an understanding of how power, influence tactics, and influence processes are manifested in virtual teams.

METHODS

Participants

Participants were 23 members of different virtual teams (See Table 3). The average age of participants was 43.64 years old, and their average tenure with their virtual teams was 2.28 years. Fifteen participants (65.22%) were female, and all participants were white. Eight participants (34.78%) were the leaders of their virtual teams, while the other fifteen participants were junior-level team members. Participants were recruited using two methods. First, participants were recruited from various groups on the online networking site *LinkedIn*. With permission of the groups' moderators, messages were posted on the groups' discussion boards requesting participants for a study on virtual teams. This was a useful recruitment strategy because it provided a wider scope of participants (i.e., cross-sector or cross-industry) and potential participants were easily accessible (Witmer, Colman, & Katzman, 1999). Second, participants were recruited through a status update on *Facebook*. The status update message encouraged people to forward the request for participation in the study to people they know who are also members of virtual teams. I continued to interview new participants until the data reached theoretical saturation.

Materials

Questionnaire

The purpose of the questionnaire was to collect important information without unnecessarily extending the length of the interview (see Appendix A). The questionnaire was distributed in an email attachment prior to the scheduled interview time. Demographic data regarding the participants and their teams is important because it provides a context for understanding the data. This data also provides information regarding the participants' status characteristics which was used to examine the effects of status on the influence process.

Interview Protocol

The interview protocol was adapted from the essay questions used by Kipnis and his colleagues (1980) with additional questions aimed at identifying the unique contributions of the technology (see Appendix B). The questions in the interview protocol dealt primarily with influence tactics because they are how people overtly exert power. Because power is an abstract construct, it was theoretically easier for participants to describe their power-related behaviors (i.e., influence tactics).

Procedure

Because this is a relatively unstudied area of research, qualitative methods were deemed appropriate as they provide rich descriptions of the phenomena under investigation. Semi-structured respondent telephone interviews were conducted in order to clarify the meaning of power and influence in virtual teams, to determine what factors determine the use of certain influence tactics on virtual team members, and to classify the influence tactics used in virtual teams (Lazarsfeld, 1944). Interviews were desirable in this study because they allowed for in-depth examinations of phenomena based on participants' interpretations of their experiences (Charmaz, 2006). Conducting interviews provided rich, detailed data necessary to build a theory of influence and status in virtual teams.

In addition to having participants describe their successful influence attempts, we also had participants describe unsuccessful attempts to influence members of their virtual team. Descriptions of failed attempts to influence virtual team members were beneficial because they showed the limitations of applying certain FtF influence tactics to the virtual environment. In addition, while it is important to identify how virtual team members can successfully influence one another using technology, it is equally important to understand what aspects of technology lead to unsuccessful influence attempts.

Interviews are time-consuming activities for participants (Jackson & Trochim, 2002; Lindlof & Taylor, 2002). To ensure participants were engaged in the study, I offered participants an incentive. Participants were entered into a drawing for one of five $50 Amazon.com gift cards.

Once individuals agreed to participate in the interviews, I emailed them a short questionnaire to collect demographic data (see Appendix A). Once participants returned the survey, I scheduled the interview. The interviews were semi-structured and designed to gain an understanding of how technology enables and constrains the availability and choice of influence tactics in virtual teams. Combined with the pre-interview questionnaire, the interview allowed me to examine how status affects the influence process in virtual teams.

Due to the geographically dispersed nature of my proposed sample, interviews were conducted over the phone. I audio recorded the interviews and transcribed the recordings verbatim. Because errors of speech and repetition were not part of the current analysis, they were removed from quotes in this paper to provide clarity for the reader.

A semi-structured interview format was chosen as it provides consistency between interviews and allows for comparisons between responses; however, because the research is exploratory, it was necessary for participants to be able to discuss issues not addressed in the interview protocol that they felt were important for an understanding of the topic.

Analysis Strategy

The transcribed interviews were imported into NVivo8 for analysis. This software was chosen because it provided an efficient way to store, organize, manage, and code the large amount of data gathered during the interview process. Theoretical memos were written throughout the data analysis process to further flesh out the thematic qualities of the coding concepts and categories.

The interview data was analyzed using a thematic approach. During open coding, the experiences of the interview participants were compared in order to uncover common influence tactics and influence processes (Corbin & Strauss, 1990). During this phase of data analysis, I let codes emerge from the data. This was accomplished primarily through the use of in vivo and process coding. These coding methods allowed me to stay true to the participants' accounts and accurately portray the actions they described. Codes and categories were revised and new categories created until all of the data was analyzed.

The data was subsequently integrated using axial coding in order to create categories and themes that span many categories. Based on previous research, I expected that traditional influence tactics and status characteristics that are present in FtF interactions would persist in the virtual environment (Elron & Vigoda-Gadot, 2006; Hollingshead, 1996; Saunders et al, 1995). As a result, during this phase of analysis certain codes became grouped as they had been in previous research of FtF teams (e.g., the in vivo code "brute force" became grouped under the traditional influence tactic of pressure). That is, certain codes of influence tactics in virtual teams began to resemble traditional influence tactics. Traditional status characteristics and status cues also emerged from the codes that were created during open coding. I then examined each construct created during axial coding and teased out key dimensions (Corbin & Strauss, 1990). Selective coding was used to unify all categories around a core category. Data continued to be collected and analyzed until the category set became theoretically saturated.

The purpose of qualitative research is not to achieve generalizability. However, generalizability can be inferred to a certain extent due to two characteristics of the methods and data. One, participants were members of virtual teams that varied in size, industry, and geographic location. Thus, sampling was somewhat reflective of the population at large, Two, theoretical saturation implies that results of the study are generalizable because consistent themes arose from the data.

In order to ensure the integrity of the categories and constructs identified during data analysis, researchers familiar with virtual teams were asked to confirm the appropriateness of the coding scheme. In order to further strengthen my interpretation of the data, I conducted negative case analysis (Lindloff & Taylor, 2002). Negative case analysis allowed me to determine if there are any instances which refute the categories I have created. When a negative case was identified, I revised my interpretation of the data and continued to compare my interpretation with new data.

RESULTS

This chapter will describe the results of the study. First, I will discuss the various categories of influence tactics used by participants. To do this I will begin by discussing the presence in virtual teams of influence tactics that were identified previously in FtF research and how these traditional influence tactics have been adapted for the virtual environment. Then I will describe newly identified influence tactics and how they are related to traditional influence tactics. Second, I will discuss the success of the influence tactics in a direction-specific manner. That is, I will discuss which upward influence tactics are successful, which lateral influence tactics are successful, and which downward influence tactics are successful. Third, I will discuss additional strategies that participants described using during their influence attempts in order to make the influence attempts successful. Fourth, I will discuss the status cues and status characteristics that are relevant in virtual teams. I will also compare status characteristics that are relevant in the virtual environment to those that are relevant in the FtF environment and discuss how status affects the influence process in virtual teams. Finally, I will discuss additional findings of this study that are indirectly related to the research questions which include communication technology preferences, culture, and the importance of time to members of virtual teams.

Influence Tactics in Virtual Teams

Influence tactics are the specific actions that people take to influence others (i.e., get a target person to perform a desired action). Two categories of influence tactics emerged from the data. These categories include traditional FtF influence tactics and a newly identified category of influence tactics: ambiguity reduction techniques. In addition, the technology allowed members of virtual teams to use traditional influence tactics in new ways.

Traditional FtF Influence Tactics Also Present in Virtual Teams

Traditional FtF influence tactics were highly prevalent in the participants' virtual teams (See Table 4). In order of their prevalence in virtual teams, these nine influence tactics are: pressure, legitimating tactics, rational persuasion, consultation, inspirational appeals, exchange, ingratiation, coalition tactics, and personal appeals. These influence tactics have been discussed at length in the literature, but rarely, if ever, in the context of a virtual team. Below I will describe the traditional FtF influence tactics as they emerged in virtual teams and how participants used technology to adapt them to the virtual environment (See Table 5).

Pressure

Pressure was the FtF influence tactic most often described in the interviews. Pressure is the use of "demands, threats, frequent checking, or persistent reminders to influence the person to do what you want" (Yukl, Guinan, & Sottolano, 1995, p.275). William[1], a junior team member, described an incident where he used pressure to get the other person to do what he wanted:

> What we basically told them is...'Listen, that's okay, but...the only funding you'll be available to get is...what you requested, which...is under what you actually need. So, if you want to tell your business partners...that you're the reason they got...a few million dollars less in funding than necessary, then that's fine, but...I don't think you want to do

---

[1] All participant names are pseudonyms.

that, so you need to get...your request filled in correctly, get it done, get it in on time, and ensure that it's up-to-date and has the right information.[2]

In the participants' accounts of their virtual influence attempts, I identified additional components of pressure: following up, frequent communication, the forwarding of previous electronic communications, guilt, and "brute force". Below I will describe these new versions of pressure and how they were made possible by the technology. Michael, a junior team member, described how he used following up to influence others:

> It was just... a matter of having to follow up...to make sure emails were seen and questions were getting answered...That kind of...at a broad level has to do with...dealing with...slow responses when you don't have the opportunity to run into somebody in the hallway or see somebody at the office every day.

Patricia, a senior team member, described how she used frequent communication to get through to her team member: "I had to keep saying over and over to him, 'The client really likes this headline. The client thinks this headline really works.'...finally he got it."

Some participants such as Richard, a senior team member, used subtle forms of pressure (e.g., forwarding previous emails) as a way to influence others: "Sometimes forwarding your last request along with the new request helps remind them in a subtle or not so subtle way that they should have already done it." It appeared that the softer forms of pressure (i.e., frequent checking and persistent reminders) were easy to use in the virtual environment given the available technologies (e.g., email).

---

[2] Ellipses in quotations indicate pauses, errors of speech, or repetition.

However, Linda and Susan, junior team members, described using more obvious forms of pressure (e.g., guilt and brute force). According to Linda, "It was usually just a matter of…'Hey, the deadline's coming up. I've got three of the four people lined up here. You're the one who's holding it up.' Guilt was a big motivator." And according to Susan, "It really takes some brute force for him to realize why this is needed… the emails have to be strong…brute force to me is more of, 'This is exactly what I need, when I need it. When will I get it?'"

One senior team member, Kimberly, and her team used what she called "Zen Mail," in which they would send email messages with the entire message in the subject line. She believed that emails with the subject beginning 'Urgent' were very influential in that they were attended to by the target immediately and could be read in their entirety without being opened. In this way, Urgent Zen Mail was related to pressure, and is an example of how communication technology allows virtual team members to enact influence tactics in new ways that are not possible when interacting FtF.

Legitimating Tactics

Legitimating tactics are the establishment of the "legitimacy of a request by claiming the authority or right to make it, or by verifying that it is consistent with organizational policies, rules, practices, or traditions" (Yukl et al, 1995, p.275). Karen, a senior team member, described using her organization's policy to influence her subordinate: "It was really just something that we're all supposed to be compliant with, and just explaining to her over and over again the reason why we all have to do this…I have to enter my hours in the system, everybody does. Even my manager does."

From the participants' accounts, I identified a new version of legitimating tactics that involved escalating the issue to a manager in order to establish the legitimacy of the request. However, as described by Michelle, a junior team member, this was often seen as a last resort when other tactics have failed: "I did have to escalate this one to two directors in my organization who are ultimately responsible for making sure that customers are supported…though it wasn't the preferred method…it did work." Similarly, Carol, a junior team member, noted that she tried to influence people on her own first but would escalate to a manager if that did not work.

Technologically, participants escalated the issue to a manager by copying managers on emails. Communication with higher level team members was easily accomplished because of communication technologies – whereas it may be difficult to get a FtF meeting with senior leadership, it is a simple enough process to send a senior leader an email. As a result, one outcome of the use of communication technologies in this way appears to be a decrease in the rigidity of the power structure in virtual teams. Susan, a junior team member, described a situation in which she ultimately had to copy her target's manager to get results:

> At first it was a reminder, you know, just a casual, even just, usually when we do it via Communicator first when we get it. So then it was a phone call. Then it was a message. Then it was a follow-up email. Never resolved anything. And it took me to copy his manager on his email yesterday in order to get it resolved today.

Participants also reported using technology to escalate the issue to a manager by forwarding previous communications to managers, as described by Helen, a junior team member: "I went back to my manager…and I forwarded her all the emails that I had been sending, and she sent an email to his manager. And I think it might be better now <laughs>."

Rational Persuasion

Rational persuasion is the use of "logical arguments and factual evidence to persuade the person that a proposal or request is practical and likely to result in the attainment of task objectives" (Yukl et al, 1995, p.275). As in the FtF environment (Kipnis et al, 1980; Yukl & Falbe, 1990), virtual team members were receptive to this influence tactic. In support of this, Daniel, a junior team member, noted, "If people understand why something is required and…there's a reason for something…then usually they'll acquiesce." Here Laura, a senior team member, expresses her opinion about influence tactics:

> I think in general persuasion is best. A logical, persuasive argument so that, you know, you just lay out the facts. You say, 'From A to B to C. And if we do that we will get to C and C is our goal. So this is what we'll do.'

Participants used a variety of technologies to adapt rational persuasion to the virtual environment. According to Kimberly, a senior team member, "We use technology to track and generate data, so that when we are making proposals we have evidence that shows realities. So we're not just relying on conjectures." Technology, in the form of highlighting digital text, was also used in order to draw team members' attention to important information. According to Sharon, a junior team member:

> If there's something that's particularly critical, like if it's a really long document, [my manager] might…point us to certain things. Like he might highlight something in yellow and say…'Be sure to look at the items highlighted in yellow.'

Highlighting important information is related to rational persuasion if the highlighted information contained factual evidence meant to persuade the target.

Consultation

Consultation is when, "you seek the person's participation in planning a strategy, activity, or change for which you desire his or her support and assistance, or you are willing to modify a request or proposal to deal with the person's concerns and suggestions" (Yukl et al, 1995). Participants frequently reported using consultation in their virtual teams as a way to engage team members in the task. Helen, a junior team member, compared and contrasted two of her team members' influence tactics, and concluded that consultation was more effective.

> I think Carol's much better at…getting the whole team involved in a decision…Julia's much more likely to just call me on the phone or Ping me, which is instant messenger…to get something done and Carol will kind of use the group's influence and will…maybe send an email to the entire team saying, 'This is what I'm thinking. Give me your thoughts.' And people tend to agree with her. That way instead of doing one-off conversations and then having to bring it to the whole team and get buy-in, she kind of does that upfront so that it causes her less problems throughout the process.

Joseph, a senior team member, believes that consultation helps him to be successful in his influence attempts:

> When we did our planning, we do it together. So people make their own commitments to certain things and times when they're doing them as opposed to me just assigning things. I think that was the biggest thing [I did to get people to do what they needed to do].

The technologies used by participants in their virtual teams were especially well suited for adapting consultation to the virtual environment. Participants reported using whiteboard technology, electronic polling, and screen sharing to encourage the participation of their team members. For example, participants used the whiteboard feature of LiveMeeting in order to collaborate with others. It allowed virtual team members to interactively share their ideas visually instead of requiring team members to focus solely on verbal or text communication. Karen, a senior team member, described how the whiteboard technology is related to consultation.

> She likes to use technology…especially the Live Meeting function that has the whiteboard feature. And she really likes to engage…her clients and team members in using that whiteboard feature to write down their ideas or suggestions…and it's a really great tool because now the client gets to see their ideas right there in front of them. So sometimes she uses the whiteboard to let them write down their suggestions, and other times she just does it live for them so that they can follow along with the process. So she's really great about using the tool for that reason.

Electronic polling was also used by participants as a way to include all members of the virtual team in the decision making process. One junior team member, Dorothy, described how her team members used electronic polling to gather information from virtual team members. She enabled the polling feature of LiveMeeting in order to involve team members' in the planning of a project.

> There's a technology that we've used a lot in Live Meetings where we take polls and votes, and that kind of thing. And that's now surfacing a lot in emails…that is just a very specific piece of technology that I didn't mention, but when we're working with groups and we need responses, we now tend to use polling or voting buttons…to gain consensus without ever having to have a meeting…it's a very efficient way to get at that answer without having to ask everybody to drop everything and get on a conference call.

In this example, polling enabled the use of consultation because it got virtual team members involved. However, polling allowed participants to engage in an indirect form of consultation – it did not require direct interaction amongst virtual team members. Electronic polling is another example of how a traditional influence tactic is used in virtual teams in ways that cannot be done in FtF interactions.

Inspirational Appeals

Inspirational appeals are "requests or proposals that arouse enthusiasm by appealing to the person's values, ideals, and aspirations, or by increasing the person's confidence that he or she would be able to carry out the request successfully" (Yukl et al, 1995). For example, Michelle, a junior team member, used inspirational appeals when interacting with her coworker: "I knew…what made him kind of excited. And so I made sure I emphasized those pieces of the project…I tried to anticipate where his resistance might lie." Robert, a senior team member and self-described "cheerleader" in his team, described how he inspired others to get the job done through a, "very positive, optimistic outlook that we can succeed…beyond that, it was a case of…communicating to people that…there is confidence from seeing that work actually gets done…communicating to people that this isn't a waste of time; we're actually getting things done."

Exchange

When you use exchange tactics, "you offer an exchange of favors, indicate willingness to reciprocate a favor at a later time, or promise the person a share of the benefits if he or she helps you accomplish a task" (Yukl et al, 1995, p.275). Kimberly, a senior team member, noted that her influence strategy, "really comes down to honesty and looking at a win-win. Trying to find what's in it for both of us so that both of us can succeed." While William, a junior team member, thought that successfully influencing someone is, "reciprocal…you do something for them…one day they'll do it for you."

Ingratiation

When you use ingratiation, "you seek to get the person in a good mood or to think favorably of you before making a request of proposal (e.g., compliment the person, act very friendly)" (Yukl et al, 1995, p.275). Ingratiation was one of the traditional influence tactics least often described by participants in the accounts of their influence attempts. Betty, a junior team member, described her manager's way of influencing others as "very cordial." Patricia, a senior team member, described how she influenced others by putting them in a good mood:

> You have to respect the other…person's contribution, and say, 'Oh wow, I see that…that's really cool. How you did that? It's cool how you make it, those slides move along.' Or, 'That's cool how you…put the headline in that way. It's really great.'

Coalition Tactics

Coalition tactics are when, "you seek the aid of others to persuade the target person to do something, or use the support of others as a reason for the target person to agree to your request" (Yukl et al, 1995, p.275). Betty, a junior team member, was the only participant to mention enlisting the help of others in an influence attempt: "I will go to either one of her direct reports…who sits down the hall from her <chuckles> and say, 'Hey, when you meet with her today, will you ask her about such-and-such?'" Although this was not stated by the participants, the virtual environment may inhibit the use of this tactic because it takes longer to establish relationships with others (Walther, 1995) and thus be able to form coalitions.

Personal Appeals

To use personal appeals, "you appeal to the person's feelings of loyalty and friendship toward you when you ask him or her to do something" (Yukl et al, 1995, p.275). Michelle, a junior team member noted the important role of relationships, even when compared to the role of the technology in her virtual team: "We manage to get work done on an informal basis amazingly well…I guess it's an outcome of personal relationships that evolve over time in an organization as much…as it is by…Microsoft Project Plan where a task is coming up." Christopher, a junior team member, also stressed the importance of friendships in virtual teams when it comes to getting what you want:

> I think [that I] have…a strong relationship, a strong rapport with that associate, when it does come time when I need a favor…I'm able…to go to that…associate and if I didn't have that strong relationship I'd have to probably devise a different strategy, but…because we're close coworkers…there's really…no need to…try to figure out how to make sure that the associate does what I ask them to do.

Emoticons were a technology feature that one participant described in connection with the personal appeals influence tactic. Emoticons allow virtual team members to share emotion through written communication. Elizabeth, a junior team member, reported that one of the reasons her manager was influential was because she used emoticons: "[The most influential person on my team is] friendly in her IM, you know, she's not afraid to use an emoticon and put a smiley face or to do something else."

In sum, participants reported frequently using traditional influence tactics in their virtual teams. They also described how they enacted these tactics in the virtual environment and how the technology made possible new versions of traditional FtF tactics. In the next section, I will describe a new category of influence tactic that emerged from participants' descriptions of their influence attempts: ambiguity reduction techniques.

Ambiguity Reduction Techniques

The set of new tactics I identified appeared to be related to reducing ambiguity (See Table 6). These techniques included information sharing, creating accountability, and giving the target an example. I classify these tactics as new not because they do not exist in FtF interactions, because they do, but because they are explicitly used in virtual teams as a way to work around the ambiguous nature of the virtual environment. These tactics were used by the participants to ensure that they got exactly what they wanted from their targets.

Sharing Information

Sharing information ensured that the target had the necessary details to complete the request. In the words of Sharon, a junior team member: "I think it's just that sharing of information that makes people feel like they want to do what needs to be done." Although sharing information is done in FtF teams, the use of technology made sharing information necessary because there were fewer cues present to clarify meaning. However, sharing information was not always easy as noted by Daniel, a junior team member at a publishing company: "That's in fact…one of the biggest problems…in virtual teams that I've run across; what's difficult is…keeping everybody informed."

Creating Accountability

Participants reported that creating accountability ensured that the target would follow through on the request. Like the tactic sharing information, creating accountability is also present in FtF teams. However, the use of creating accountability is more explicit in virtual interactions due to the written record. According to Betty, junior team member, these tactics also reduced any confusion that could exist surrounding the influence request:

> She also…has the…very quick ability to hold people accountable…if you say, 'Well I think it'd be a good idea if we brought cookies to the party.'

She would immediately say, "Will you take responsibility for that?' So, you know, 'And if not you, then can you find someone on your team? Okay, I'll put you down as the one that's bringing. And how many cookies will you bring? Great. And will you be there any earlier? Do you need, do you need a plate for the cookies?'…I think …influence at the moment, but I also think it's about influence…over time. In other words…we're not gonna have to revisit this. Nobody's questioning who's bringing the cookies…it was said on that call and we all heard it. It'll show up in the minutes or on the project plan that way. So the influence is in that moment to get the information, but then it's also the follow-through.

Giving Examples

Giving examples of what the agent wanted also helped ensure that the target produced exactly what was requested. Carol, a junior team member, described a time she gave a target an example to help get what she wanted: "Our plan is almost usually…we reach out…I showed him what we did with Randy. I showed him…what format Randy sent us in…so he knew what we were looking for." Giving examples is another example of an influence tactic that is used in FtF interactions but is made explicit in virtual teams because the technology makes it more obvious. Participants gave examples easily by using email and screen sharing.

Sharing screens was one way participants reported using technology to give their targets examples of what they wanted. When participants used screen sharing technologies, the target could see what was being referenced during a phone conversation. Elizabeth, a junior team member, described how screen sharing provided additional clarity to requests:

> We just used Communicator again, where I shared my screen and so I walked him through, 'Well this is how an item or a hyperlink is displayed on the Excel document and this is what I need, so um, whatever this title is in the Excel document, this is what it's referencing,' and showing him on the SharePoint[3] site the area and the section that I was trying to map it over to.

Summary

Virtual team members used a wide variety of influence tactics to get their team members to do what they wanted. These influence tactics ranged from those identified previously in FtF teams to ones that were necessitated by the virtual environment (e.g., ambiguity reduction techniques). Below I will discuss how the use of these influence tactics varied depending on the direction of the influence attempt (e.g., lateral, upward, or downward).

Directional Differences in the Success of Influence Tactics in Virtual Teams

The previous description of influence tactics provided a broad picture of what influence looks like in virtual teams. Influence tactics were not limited in their use to a specific direction of influence; however, the influence tactics that were reported by participants were not equally successful for each direction. Tables 7-9 provide a summary of successful and unsuccessful tactics by the direction of the influence attempt. Upward and downward influence tactics were, for the majority of cases, successful. However,

---

[3] SharePoint is a Microsoft product. SharePoint sites provide a single infrastructure for all of a company's websites. SharePoint allows employees to share documents with colleagues, manage projects with partners, and publish information to customers. SharePoint can also be used as a web application development platform.

lateral tactics were only successful about half of the time. Below, I will provide a more detailed description of the influence tactics available in the virtual environment in order to answer: How do low-status individuals successfully influence other members of their virtual teams?

Upward Influence Tactics

Being the low-status person in the influence interaction, in terms of occupational position, did not limit the influence tactics available to participants (See Table 7). In spite of not being limited in their choice of influence tactic, over half of participants reported using traditional influence tactics in the upward direction. In addition, the majority of participants who described upward influence tactics reported that their influence attempts were successful.

The most frequently used influence tactic in the upward direction – pressure – was most often successful. However, previous FtF research has shown that assertiveness and pressure are used infrequently, particularly in the upward direction (Kipnis et al, 1980; Yukl & Falbe, 1990). The fact that the virtual environment enabled the use of more aggressive influence tactics helped lower status team members be successful most likely because it helped low-status members to be heard by their targets.

In addition, rational persuasion was successful each time participants reported using it in the upward direction. The success of rational persuasion echoes results from previous FtF research which showed that subtler influence tactics (e.g., consultation and rational persuasion) are used more frequently and are more successful in FtF interactions (Yukl & Falbe, 1990). Because of the pervasive success of this tactic in the FtF environment, it makes sense that it would also be successful in the virtual environment.

Interestingly, no participants reported using ambiguity reduction techniques in the upward direction to influence their targets. One explanation for this could be that low-status team members were more concerned that their influence attempt was received by the higher status target than that it would be misinterpreted.

Why would someone allow themselves to be influenced by a team member of lower status? One potential explanation for the success of upward influence tactics comes from the task-oriented nature of virtual teams. This focus on the task, rather than on the team structure, might have given more power to low-status members. Team leaders may have been more willing to listen to low-status members given that the teams consisted of experts and the leaders wanted to accomplish the task effectively and efficiently.

In summary, pressure is a tactic that was frequently used and highly successful in the upward direction. Based on participant accounts, I conclude that low-status members need to be assertive in order to be heard in the virtual environment. They need to be clear as to why people should listen to them.

Lateral Influence Tactics

Participants reported the successful use of influence tactics in the lateral direction a little more than half of the time (See Table 8). Two of the most successful lateral influence tactics were traditional FtF tactics: exchange and pressure. One explanation for the success of exchange tactics could be that they were occurring between status equals. Participants may have been more willing to help their peers if they knew they could count on them when they needed help. Interestingly, in previous research on influence in FtF interactions, exchange and pressure were the two least used tactics in the lateral direction

(Yukl & Falbe, 1990). Instead, consultation and rational persuasion were the most frequently reported influence tactics in the lateral direction.

When discussing lateral interactions, many participants reported that if they had had the opportunity to interact FtF, they would have taken it; however, they didn't feel that it would have affected how they approached the target or how the target would have responded. Others felt that it would have been beneficial to have the ability to read the target's facial cues or body language or to check the target's availability. They also felt that there were fewer communication issues and the influence process could be accomplished more quickly when people interacted FtF.

Downward Influence Tactics

Downward influence tactics were reported to be successful a majority of the time (See Table 9). Pressure and rational persuasion were the most frequently used and the most successful tactics in the downward direction. These results are similar to those of previous FtF research which found that consultation and rational persuasion were the most frequently reported successful influence tactics (Yukl & Falbe, 1990).

Interestingly, participants reported using a wider variety of ambiguity reduction techniques in the downward direction than in the lateral or upward direction (ambiguity reduction techniques were not reported in the upward direction). Supervisors have structural authority which increases the likelihood that their influence attempt will be successful. This may be one reason why their focus was instead on making sure that their target properly understood the request.

Overall Analysis

The availability of influence tactics was not contingent upon the direction of the influence attempt (e.g., upward, downward, or lateral). Participants described the use of both traditional influence tactics as well as new influence tactics that have emerged due to the virtual environment and the technologies that are associated with it (e.g., adaptations of traditional influence tactics and ambiguity reduction techniques). Thus, all members of virtual teams have ample opportunity to use a variety of influence tactics. However, no participants reported using ambiguity reduction techniques in the upward direction.

Participants were more likely to use traditional influence tactics instead of ambiguity reduction techniques. One explanation for this is that past behavior is the strongest predictor of future behavior (Ouelette & Wood, 1998). The majority of people who are currently in the workforce did not grow up with the technology that is available today, so they are not accustomed to using it to influence their coworkers. Because it is what they have always done, members of virtual teams still rely heavily on the influence tactics they use in FtF interactions; however, the virtual environment has provided people with the opportunity to engage in these tactics in new ways. For example, if an influence attempt was not proceeding as planned, participants would frequently escalate the situation to their managers (i.e., using their manager's authority to establish the legitimacy of the request). Technology allowed this to be done easily by copying a manager on an email or forwarding previous emails to a manager.

Although there is reliance on traditional influence tactics, findings from the current study differ from studies of FtF influence tactics. The three most frequently used influence tactics in virtual teams were the use of pressure, legitimating tactics, and rational persuasion. This finding is not in agreement with previous research on FtF influence tactics in which the most frequently used tactics were consultation, rational persuasion, and inspirational appeals (Yukl & Falbe, 1990).

Pressure was a commonly used influence tactic in the virtual teams, but this 'hard' (i.e., more obvious and forceful) influence tactic is not as successful when enacted FtF (Yukl & Falbe, 1990). The virtual environment necessitated the need for tactics such as pressure because influence attempts are easy to ignore in this context, as described by Michelle, a junior team member: "The virtual tools have their limits in that people can avoid them better <chuckles>, so if I'm relying totally on telephone and voicemail and email and scheduling appointments…it's easier to duck that than it is somebody standing FtF." The online disinhibition effect explain how factors of the virtual environment, such as asynchronicity and the minimization of authority, empower people and explains how they are able to and need to be more assertive than they normally would be in FtF interactions (Suler, 2004). In addition, computer-mediated communication has been proposed to free people from conforming to social expectations (Dubrovsky et al, 1991).

The use of more assertive influence tactics could also be attributed to the ambiguous nature of the virtual environment. Being left in a state of uncertainty (e.g., *Is she ignoring me or did she not get my email?*) may have prompted many members of virtual teams to use various forms of pressure, including following up, frequent communication, and forwarding previous communications as reminders, to ensure that their influence request was heard and that their target acquiesced. Ambiguity reduction techniques were also ways for virtual team members to get their way.

Additional Strategies Used in the Influence Process

In spite of their reliance on traditional influence tactics, people have adapted to their new virtual environments and, as a result, a new influence process has emerged. In addition to ambiguity reduction techniques, the introduction of a wide variety of communication technologies also enabled participants to develop a strategic use of technology to get attention. In the following sections I will describe this and other strategies that participants used to increase the likelihood of a successful influence attempt.

Strategic Use of Technology to Get Attention

One way in which the influence process in virtual teams differed from the process in FtF teams is that it is much more difficult to get the target's attention in the virtual environment. According to Linda, a junior team member, "the emails, the voicemails, the other messages, it's all…the same source. It's all the same message…he couldn't have ignored me in person as much as he ignored my emails and voicemails." Thus, getting the target's attention was a crucial first step in the participants' descriptions of their influence attempts. Participants' accounts of their influence attempts illustrated how they strategically chose communication technologies in order to get the attention of their target. Below I will describe the reasons why participants chose to use certain media to get their targets' attention and how it could increase the likelihood of a successful influence attempt (See Table 10).

Multiple Communications

Participants often used more than one technology to communicate with their targets. They offered two reasons for the use of multiple communications. The first reason was preventative: participants wanted to ensure that their request was heard. For example, multiple communications gave Susan, a junior team member, additional access to her target: "One of our vendors that we're working with right now [travels] all the time...So we've had to use several different modes of communication to get to them." The second reason participants gave for the use of multiple communications reactive: when participants were unsuccessful in their first influence attempt, they moved to another media so that their request could not be ignored as easily. William, a junior team member, described a time when he switched from email to phone in order to convey his influence attempt more effectively:

> At first I was relying on emails...which just wasn't getting the job done....
> So I moved to a conference call. Kind of expressed...how we needed to
> get it done with my tone of voice...just really clearly explaining the
> deadline's in place and...the correct way to...complete what we were
> trying to get done.

Similarly, Dorothy, a junior team member, changed her communication technology when her first attempt was ignored by the target: "I emailed him...first. And then, actually when I needed a response and had not heard from him...I did watch to see if he was available for an instant message and...went in that way as a second attempt." Either way, the use of multiple communications increased the likelihood of success.

Typically, participants chose to email their targets first and then follow up with a different communication technology. Sharon, a junior team member described her communication strategy:

If I'm dealing with individuals...more often than not I will email *and* IM them. Especially if it's something that is... really important and I need a fast turnaround...because I want to make sure that...the email doesn't get lost and they haven't seen it...for my boss and others I will definitely IM them...sometimes I just IM them to say, 'Hey, no rush. I just sent you a request. No rush, just want to make sure you saw it, but I need it by such-and-such a date.' [italics added].

Use the Target's Technology of Choice

As another means of ensuring their influence attempts were not ignored and to make the odds of success more likely, participants often chose to use their target's technology of choice. Participants reported that their targets were more likely to listen to and comply with the request if it was communicated using their technology of choice. Targets were also less likely to ignore communications if they came via their preferred medium. An example given by Donna, a junior team member, illustrated this well:

I always use her preferred form of communication first, and then I'll slowly introduce the other form to get what she needs...So for example...she needed some information from me so she used her form of communication, which she prefers to instant message and email first...I used that with her, and then with my answer I directed her to a SharePoint site that had all the information in it. So I still used her preferred form of communication to establish that trust with her...that we're on the same page. I hear her, I understood what she wanted. And then when it came to giving her the answer that she needed from me, I gave her the SharePoint site which had all the details on it plus more information than she wanted, but it kind of give her an introduction of what a SharePoint site would be able to…facilitate for her particular program.

Use of Instant Messaging to Check Availability

Another tactic that participants reported using in order to get their targets' attention was the use instant messaging (IM) technology to check the availability of the target before a request was made, thereby ensuring that the request was seen. Michael, a junior team member, described how he used IM: "I usually use Communicator before I call. So I'll just send a quick message saying, 'Is it okay if I call right now?' before making that phone call." The use of IM was a strategic choice participants made to get their targets' attention. The technology allowed them to do what people do when they work FtF, namely look in someone's office to check their availability. Like using the target's technology of choice, these IM tactics helped the participants ensure that their targets would not ignore their requests.

Summary

The use of multiple communication technologies, using the target's technology of choice, and using IM to check the target's availability were all ways that participants used technology to get their targets' attention. The majority of influence attempts were successful when they incorporated one of these tactics to get the target's attention. However, getting the target's attention was just the one strategy participants employed to ensure their success. Below I will describe how participants also used strategies related to relationship building to increase the likelihood of successfully influencing their targets.

Influence Strategies in Virtual Teams

In participants' descriptions, influence manifested itself in the form of tactics and strategies. An influence strategy is a more global approach to influencing someone (e.g., I want an inclusive team) than an influence tactic. The influence strategies described by participants were closely related to influence tactics in that they affected whether or not attempts to get another member of the virtual team to do something the actor wanted were effective.

Two categories of influence strategy emerged from the data: relationship building (See Table 11) and documenting communications. Relationship building is the establishment of personal connections with virtual teammates. Building relationships, or one of its components, was mentioned by 16 of the 23 participants. Documenting communications is a strategy in which members of virtual teams keep written records of communications with their team members.

Influence strategies increased the likelihood that influence tactics were successful and facilitated the use of certain influence tactics, such as exchange and personal appeals. It is something that participants considered to be a best practice, a strategy that they tried to use at all times. It is interesting to note that, for each incident in which an influence strategy was explicitly used, the participant was successful.

Building Relationships

Building relationships was not a part of any of the reported unsuccessful influence attempts. In the majority of instances, building relationships was discussed in general terms by participants. Michelle, a junior team member, expressed her need for personal relationships in the workplace:

So, speaking personally I've been working here for almost thirty years and there are several people in that same boat. So we manage to get work done on an informal basis amazingly well. And it's, um, I guess it's an outcome of personal relationships that evolve over time in an organization as much as it, as it is by, you know, Microsoft Project Plan where a task is coming up.

The components of the relationship building process are discussed in more detail below.

Getting to know others & putting a face with a name. Getting to know others is a critical first step in establishing relationships. As with the other components of building relationships, getting to know others was reported by participants to take longer in virtual interactions than in FtF interactions. As stated below by Charles, a junior team member, putting a face with a name was important because participants felt that meeting FtF expedited the process of getting to know others.

I think it's just a slower process of having to get to know people, and what's…been very helpful for me is the fact that I have gotten to meet two of the people…from my team FtF…I've met the project manager. I've worked with her for a couple of days in person…I also worked with…the other guy for…two weeks in person…that was good because…I think it's just helpful…it sounds kind of stupid, but just having a face to go with the name when you're talking to people and when you're emailing with them…to kind of get a little bit better sense of…who they are and how they might respond to different things. So, you know, after meeting the project manager in person, I came to realize how…she's a funny…down-to-earth person who's just trying to do a good job, and so it really made it even easier afterward to…be much more open with my communication style with her…which I don't think would have happened as quickly had we not met and spent some time together in person.

Establishing trust. Establishing trust emerged as a second important step of building relationships. Trust not only affected who people chose to influence, but it also affected whether or not the influence attempt was successful. Influential team members were able to inspire a great deal of trust from their team members and were able to leverage this trust into power and influence. Dorothy, a junior team member, described the results of her influence attempt: "She approved that request for me…that seems minimal but it was…important and…it took some time for me to build some trust and credibility with her for her to just do that without any questions."

Team building & building a sense of community. Finally, team building is an important component of any team. It helps maintain relationships, which is especially critical in virtual teams. Team building has been associated with important outcomes, such as improved team functioning in cognitive, affective, and process outcomes (Klein et al, 2009). Karen, a senior team member, described how her team incorporated technology into their team building:

> We use [the digital whiteboard] a lot for team building…for example yesterday we had just kind of a fun kick off question which was, you know, 'What is your, you know, favorite carnival ride?' But instead of writing the word, we had everybody draw a picture of it. So it was just…a fun little thing that we do for about the first five minutes of our team

meeting, just to…get each other, you know, a little bit more familiar with each other, and just kind of make it a little bit fun…It's really important…because…none of us are FtF, so it's important that we do little activities like that. That might sound silly to other people. We probably wouldn't do it in <chuckles> a FtF meeting, but because it's virtual, you know, we really feel a need to have something that connects us.

Sense of community was related to the interpersonal relations component of team building. According to Robert, a senior team member: "My job was to build that sense of community that we are a community of experts."

In this way, a process for building relationships emerged from the data. First participants got to know their team members, which then allowed them to establish trust with them. The final phase of the process was ongoing – participants had to work to maintain the relationships. Building relationships was a strategy that enabled the use of the personal appeals influence tactic, which involved appealing to a person's sense of friendship.

While participants were quick to note that building relationships was an important component of working on any team, FtF or virtual, they also made it clear that the virtual environment made it more difficult to establish those relationships. Kimberly, a senior team member, noted: "I think when you can have a FtF engagement [establishing relationships] can be done faster, whereas if you're totally dependent on…different forms of virtual…relationships could take a little longer."

In sum, building relationships was critical to the success of influence tactics. An influence attempt is meaningless if it comes from someone you do not know.

Christopher, a junior team member, described the importance of building relationship in the following way:

> I think you've got to start by building a relationship earlier and a rapport. So the fact that, you know, over the years I've built a relationship with this associate...I think is critical...when you need something in a timely manner, whether you're working virtually or in person.

Building relationships with virtual team members through the establishment of trust and getting to know others was associated with successful influence attempts.

Documenting Communications

Documenting communications is an influence strategy best defined by the colorful emic code: 'cover your ass' email trail was a colorful emic code. This code demonstrated one way that participants were able to create accountability by using technology. Documenting communications is one enactment of an influence strategy that is new in virtual teams and not possible in FtF interactions. Email trails provided, among other things, a record of what requests were made, when they were made, when the target agreed to comply, and what deadline was agreed upon by both parties.

Documenting communications was similar to the creating accountability influence tactic in that written records could be used by participants to create accountability with the target. However, documenting communications was categorized as a strategy, rather than a tactic, because it was an ongoing pattern of behavior for participants. Email trails were not necessarily kept for the sole purpose of influencing team members.

Influential members of virtual teams, such as Sandra (a senior team member), kept records of their interactions so that they could go back and use them, if necessary, to ensure that their influence attempt was a success.

> I document everything electronically. And that's the easiest way for me to do it...have something in writing saying, 'This is how we contracted for this'...not so much as if somebody doesn't do something, but that everybody has a reference to go back to.

Summary

The virtual environment creates additional difficulties for influence attempts. Participants reported using a variety of strategies and technologies to adapt to the virtual environment. These strategies including using communication technologies in various ways to get their targets attention as well as building and maintaining relationships with their targets. Participants who reported using these strategies were more likely to report the influence attempt as being successful than unsuccessful.

In the previous sections, I have discussed and analyzed the influence process as it was described by participants. The influence process in virtual teams includes not only the use of influence tactics but also the use of a variety of "pre-influence" strategies. In the following sections, I will discuss the status cues and status characteristics that were mentioned by participants. I will then describe how status and influence are related to one another in the context of virtual teams.

Status Characteristics Theory in Virtual Teams

Status issues emerged from the data through participants' descriptions of their team members. The status characteristics that participants described as differentiating their team members did not reflect traditional status differences such as gender and race. Participants relayed information about status through their description of status cues and status characteristics.

Status Cues

Status cues are "verbal and nonverbal social cues that help actors form expectations" (Wagner & Berger, 2002, p.56-57). Participants described several status cues that were relevant to their virtual interactions with team members, although status cues were not as pervasive in virtual teams as they are in FtF teams (See Table 12). In fact, status cues were only mentioned by five participants. This finding corresponds with Table 2 which shows that the majority of traditional status cues are not available when using communication technologies.

The amount a team member talked and whether or not they took initiative indicated their participation level in the virtual team. Helen and Dorothy, junior team members, described how high status members spoke up and initiated communications.

> Over the phone and over like a conference call I think it's your style of how you talk. So part of it's the amount you talk 'cause sometimes I won't even realize that someone's on a call because they didn't say anything. (Helen)

> I have been virtual for about ten years myself, and I find that I'm successful because I am out there initiating…the communications, initiating meetings, initiating, um, schedules and what needs to happen next, and just staying on top of things so that people never wonder where I am or why they haven't heard from me. And then I have seen others who

work virtually, and sometimes you have no idea what they're doing all day long. Because they're not taking the initiative to communicate, to follow up, to plan, to involve people, to engage in conversation, to schedule meetings. You really have to take that initiative, and that is, that is the, a key distinction. (Dorothy)

One junior team member, Helen, noted that a team member's voice provided cues to the person's age, and people made assumptions based upon these cues, even if these assumptions were inaccurate.

I think your voice and how old you sound makes a big difference. So, um, there was two members of our team, same position, and one sounded so much younger than the other that it really influenced the way that I thought about them and how much seniority I thought they had or how much, you know, authority I thought they had in this situation. But then when I met them it turns out they're opposite ages than what I thought. Their voices just sound differently. And it changed my entire impression of them. So that's interesting. So I always think about that because I know that I have a really young-sounding voice, so when I'm on the phone with people and I know they're never going to meet me, I always wonder how much that's influencing our discussion and how much influencing power I'm going to have because I sound so young.

Related to a team member's voice was the volume at which they spoke. Team members who spoke too loudly or who interrupted were described as lower status by

Helen: "There are other people who are loud or interrupt a lot and I think that people notice that and they lose some credibility."

Participants such as Michelle, a junior team member, expressed a desire for nonverbal cues, such as body language, in the virtual environment:

Well something, yeah, something I've learned, well I miss body language. That's the most the miss, or the biggest thing I miss about people actually being physically collocated in the same location. And I didn't realize how important it was until we started doing more and more work virtually.

The participants described fewer status cues than are present in the FtF environment. This is most likely due to the lack of information provided in the virtual environment compared to the richness of FtF interactions. For those participants who did mention them, status cues played an important role in their interactions with team members. I will now describe the various status characteristics that emerged during data analysis.

Status Characteristics

Two categories of status characteristics emerged from the descriptions of virtual team members. The first category of status characteristics consisted of diffuse status characteristics traditionally found in the FtF environment (e.g., race or sex) (See Table 13). The second category of status characteristics consisted of specific status characteristics. Many of these have traditionally been found in the FtF environment, but I identified some specific status characteristics that are only relevant in the virtual environment (See Table 14).

Diffuse Status Characteristics

Although the participants did not feel they were as relevant in the virtual environment as they are in FtF environments, several diffuse status characteristics were still mentioned by participants. These included age, occupational position, sex, race and ethnicity, parenthood, and education. Age was the most frequently mentioned status characteristic in virtual teams. In virtual teams, age was considered high status when participants described age in terms of the organizational tenure of their team members. This is because experience and knowledge of the business was a critical component of organizational tenure. Michael, a junior team member, in his team, explained why occupational position affected interactions even though it was not supposed to do so:

> No, I wouldn't say [organizational rank is] supposed to play a role [in the virtual team]. But I think people default to that…I think it's just natural that the younger people on the team are probably going to tend to default to the older people on the team…the older people on the team are gonna typically have more experience than the younger.

However, older was not always considered to be better in virtual teams. Differences in comfort with technology were related to age, and participants illustrated the potential drawbacks of being older and working in the virtual environment. Team members were described as low-status if they were uncomfortable with the available technology or did not know how to use it. They were also described as low-status if they were threatened by the new technology and were resistant to change. Robert, a senior team member, stated: "The technology threatens that whole…process gate-keeper kind of mentality." The characteristics of organizational tenure and comfort with technology were described in relation to team members' ages.

Occupational position was a relatively clear cut status characteristic in terms of who was considered to be high-status and who was considered to be low-status. Status differences were associated with a person's position based on either their profession (e.g., consultant) or their organizational position (e.g., Vice President of Marketing). Linda, a junior team member, described her team member in the following terms: "He's a very high-level physician. World-renowned…very productive, very smart, very high…ego level." Charles, a junior team member in the U.S. military describes his team member in the following terms: "I would say that the project manager has a lot of influence due to the nature of her position…because…she's the person driving the train, and she's the highest ranking person there." Those in a high-ranking profession (e.g., physician) or who were higher up in the organizational hierarchy (e.g., Chief of Surgery) were high-status members of the virtual team.

Some participants demonstrated an awareness of their team members' sex and race/ethnicity: In describing the differences between himself and his manager, Michael, a junior team member, stated: "Can you tell me a little bit more what you're looking for? I mean, I can give you the obvious, obvious things like, uh, you know, my manager's a woman. I'm not a woman <chuckles>." Kimberly, a senior team member, noted: "We've had Europeans and Americans and Asian members all on the team. And it really did reveal different working styles." Nonetheless, the participants who mentioned their team members' sex and race/ethnicity generally did not think that these diffuse status characteristics were relevant in their team and that they did not impact their interactions. As Laura, a senior team member, expressed:

> I think that the gender and the ethnic stuff doesn't really affect anybody too too much. Um, we're actually fairly evenly split men and women, so I don't think that's a big deal. Um, I think we're, oh, I have to think about racial group everybody's in, um, because *I don't really think of them that way.* Uh, one, two, three, four, four of us are Caucasian. And six of us are African American. So not that diverse actually <chuckles> we've only got two, two options. Um, but that doesn't seem to have an effect on anything either [italics added].

This quote shows that participants consciously viewed their team members in ways other than by gender and race. Thus, any effects of these traditional status characteristics were implicit. It is possible that participants' behaviors were affected by their own or their team members' gender or race; however, participants were unaware of these effects. The effects of diffuse status characteristics were not as important as other relevant, specific status characteristics, as will be discussed later in this section.

One participant illustrated how gender differences do, in fact, still affect virtual interactions. Below, Michelle, a junior team member, described how communication styles can vary based on the sex of the person.

> Male and female communications are very different…Men tend to be more assertive. So, when a male engineer is speaking to a group of people, the tendency is to say things with assertion that even, even though that person may be asking a question. What I find with women is sometimes quite differently, or quite different. A woman will say something assertively in the form of a question…A woman says, wants somebody to

shut the lights off before a presentation that's taking place. So she'll say, 'Would somebody mind getting the lights?' Which is really an order asked in the form of a question. Whereas a male would say, 'Somebody get the lights'... I tend to be, um, looking for it in the sense to say, um, 'Did you really mean that as a question or were you asking me to do something?'

In sum, members of virtual teams were aware of their team members' biological and demographic characteristics such as sex and race/ethnicity. On the whole, differences in these status characteristics were not considered by members of virtual teams to create status differences. However, from the quote above, it can be seen that men were able to maintain their traditional dominant role, even if it was not a common occurrence or frequently went unnoticed in the virtual environment.

Parenthood was also mentioned by participants as a characteristic that distinguishes team members from one another. Although participants tended to mention parenthood when describing female team members, the characteristic of having children was also mentioned by Laura, a senior team member, when she described her male team member:

He just had, his wife had a new baby in March and they have a two year old already, and I think...pressures, not pressures at home, but just having two little kids at home, he perhaps didn't focus as much as he needed to at work.

While parenthood was often simply used as a descriptor (without a positive or negative connotation), team members with children were described by some participants as not as focused at work or having limited availability at certain times of the day. Thus, when activated as a status characteristic, parenthood most likely reflected lower status in the context of virtual teams.

Participants demonstrated knowledge of their team members' education level (i.e., highest degree attained). Richard, a senior team member, stated: "I'd say [we're] dissimilar in, I guess I went to college and he didn't go to college." When describing his team member, Charles, a junior team member, stated: "He has a master's degree in I/O Psychology and he has an MBA and…so his education is much closer to mine, so we speak on…the same…level on a lot of things." Having more years of formal education is traditionally associated with a higher status. As will be discussed in further detail in the section on specific status characteristics, knowledge is extremely valued in virtual teams. Thus, being well educated is associated with higher status.

Time zone differences could also reflect status differences at the team level. As noted by Betty, a junior team member: "A lot of our…folks especially in technology are in California, so we always tease that we are…superior…everything's on Eastern Time, right? ...waking them up for their 6am and our 9am is never a fun thing." Thus, individuals in the same time zone as their organizations' headquarters were considered higher status.

In sum, participants identified several diffuse characteristics in their virtual team members. However, their descriptions of these characteristics did not typically reveal an awareness of associated status differences. Instead, status differences were revealed primarily in participants' descriptions of specific status characteristics (e.g., skills and knowledge).

Specific Status Characteristics

The second category of status characteristics consisted of specific status characteristics that revolved around skills and knowledge. Specific status characteristics were more prevalent in virtual teams than diffuse status characteristics. Skills, in general, and effective communication skills, in particular, were the most frequently mentioned status characteristics. Knowledge and expertise were also status characteristics that were often mentioned by participants as part of their descriptions of team members. Tech savviness also emerged as a new and important characteristic to possess as a member of a virtual team.

Participants described their team members most frequently in terms of their communication skills. This was especially the case when participants described the most influential members of their teams. Out of these descriptions emerged characteristics of effective communication. As described by Christopher, a junior team member, the most important component of effective communication in virtual teams was being direct and assertive:

> Well, I think [she gets other people on the team to do what she wants through] direct communication. So I think laying out exactly what the needs are for an associate to complete and…providing direction on it. So…I wouldn't say anything, anything other than…a direct style of communication

Effective communicators also provided clarity and were specific. According to William, a junior team member:

> The most important thing that [my most influential team member] provides is a sense of…organization, a sense of…clear deadlines…a clear

direction of what needs to be accomplished that often you don't get…on virtual teams, and you know as I've said, there's…a lot to be left to interpretation oftentimes at the conclusion of virtual call, and he doesn't leave it to interpretation.

In the words of Carol, a junior team member:

> When it comes to communication on things…especially working virtually, I think you have to be more specific. You know, 'I need it by this date,' you know, so we, you can turn around, turn it around by another date. So…we try to plan ahead…and know like, 'Okay, well if we need it finished by Friday, then you need to give me the information by Wednesday.'

Effective communicators used their tone of voice to help convey their message, as described by Charles, a junior team member:

> She does a good of job of kind of…conveying, you know, excitement and tone and things like that through both her voice and…when she's, you know, writing emails and such…which I think, makes her communication a little bit more real and…you can see her personality coming through in [her emails] pretty well.

As expressed by Betty, a junior team member, below, effective communicators chose their words carefully and asked questions.

> I think how we're different is, I call her, um, in my head, probably just one of the best politicians…things that are just a negative <unintelligible> and she just crafts them to be like the best opportunity <chuckles>…she

> interprets things for other people out loud…she tends to present and, and I think influence by interpreting it for you. Telling you what it means for you and how you should feel about it…And the reason she's so influential is because she asks really good questions…she asks the right questions at the right time, um, and even if she knows the answer she, she asks questions in such a way so that people have to name what's there.

Finally, effective communicators controlled their emotions, listened to others, and controlled the pace of the conversation. Christopher, a junior team member, described the importance of keeping emotions in check when communicating through technology:

> I think the other one is that…emotionally…there isn't involvement. So taking out some sort of emotion from, uh, either reading too much into…a communication or an email…that I think people think may have emotion but really doesn't…so I think being able to keep emotions in check or…in line with what's going on from a communication standpoint or just an overall interaction standpoint. I think those are probably the two biggest things that…help make for a successful virtual team.

Betty noted the importance of listening to others: "I would say…just the influence…that one uses by using coaching language or…listening skills…empathy…things like that, to be able to, I think be a more…powerful…influencer." And William, a junior team member, described setting the pace of conversations: "Well what differentiates is…some people try to talk too fast…on these calls…and kind of get lost whereas others kind of, they pause for questions."

Effective communication was important to participants because it helped avoid and clear up misunderstandings. The potential for misunderstandings can be greater in virtual teams due to the lack of cues. As one junior team member, Daniel, described: "The misunderstandings that can sometimes come up from email, you know, where sometimes…you hit the send button a little bit too soon."

Effective communication made members of virtual teams more influential because they could get the attention of others. Participants frequently described their influence attempt as being ignored by their target. As Michelle stated: "The virtual tools have their limits in that people can avoid them better <chuckles>, so if I'm relying totally on telephone and voicemail and email and scheduling appointments…it's easier to duck that than it is somebody standing FtF." According to Michael, a junior team member at a large banking corporation, email was especially easy to ignore:

> Sometimes you can send an email and the person on the other end of that email might just ignore it if they don't know who you are. And in a company with 280,000 people, there's going to be a lot of people who don't know who you are.

However, when members of virtual teams communicated effectively, people listened to them. Their team members understood what they wanted, when they wanted it, and how they wanted it done. They were, therefore, more likely to get what they wanted. The way the influence attempt was communicated to the target determined whether it would be successful, more so for members of virtual teams than for members of FtF teams because of limited status cues in the virtual environment.

Effective communication was not the only skill that was relevant in virtual teams. Participants also spoke about their team members in terms of skills which gave them the capacity to complete the tasks at hand. For example Sandra, a senior team member, said: "I don't feel that she was placed in the right type of, you know, the right project role for that project. Um, with her skill set and things like that." Joseph, a senior team member, described his team member in the following terms: "His particular skill set is fairly unique and so it's been very important for the quality of the product we're putting out." If participants were given a choice as to who to target in their influence attempts, they often chose team members who had the skills necessary to complete the requests. Thus, possession of task-relevant skills reflected high status in virtual teams.

Participants also frequently spoke of their team members in terms of their knowledge and expertise relevant to the task. Compared to knowledge, expertise reflected a deeper understanding of a topic. According to Donna, a junior team member, her team member has: "a huge…knowledge of what's going on…in her area…and she has a lot of, um, history with a lot of the other managers in the lines of businesses." Laura, a senior team member, described her coworker in the following terms:

> He's knowledgeable of, you know, what we've done before, how come we did it, why aren't we doing this anymore…and he knows the people to call within the organization to get either information or motion on different things.

Similarly, Charles, a junior team member, said:

> I'd say his influence primarily comes through…his expertise…I think it's basically, you know, through telling these…stories and talking about

different ways that he's…done things elsewhere or seen things work that really…conveys that expertise and allows him to be influential.

Knowledge and expertise were highly valued in virtual teams because they enabled participants to complete the task at hand. Thus, the possession of knowledge and expertise were high-status characteristics in virtual teams.

Tech savviness is a specific status characteristic that is unique to the virtual environment. In order to successfully work on a virtual team, members needed to be able to use the available technology, otherwise they would not be able to complete the task or communicate with their team members. Tech savvy-ness is defined by a certain type of knowledge and skills; however, this code differed from skills, knowledge, and expertise in that tech savviness was perceived by participants in terms of frequency. That is, those team members who most frequently used the available technologies were considered to be tech savvy, regardless of whether or not this reflected any actual skill or knowledge. Examples of tech savviness are given by junior team members Elizabeth and Carol:

> I know my perception has been, how have they been able to go this long without being comfortable in those tools? Because for me those are two tools that I use, as I said, every day…and so for me they are a necessity and, I mean, it's just like if someone…has a hard time copying and pasting within [Microsoft] Word. I mean, it seems to me like it's a very basic…level of comfort that you need to have to be able to be successful in the organization, and so I've usually been a little…surprised when I've come across people who don't know how to use those tools. (Elizabeth)

> I think some people really utilize the technology for the virtual teams...I'm not afraid to go out and IM somebody or, um, share my screen, I think some people really utilize their tools and others not as much. (Carol)

Virtual team members who were not adept at the available technology were considered low status.

Summary

Diffuse and specific status characteristics are available in both the FtF environment and the virtual environment. However, the emphasis on which status characteristics were most important shifted in the virtual environment. Together, the limited identity cues—compared to those available in the FtF environment – and the task-focused nature of virtual teams made specific status characteristics more important in the virtual environment than diffuse status characteristics.

How Status Affects the Influence Process in Virtual Teams

Status affected the influence process in virtual teams through status characteristics and status cues. In other words, status affected who was influential in virtual teams. Age and occupational position were the primary diffuse status characteristics that were recognized by virtual team members. They imparted influence to older team members and those who were higher in either the organizational hierarchy or who had a more highly esteemed profession.

Skills, knowledge and expertise were specific status characteristics that determined the distribution of influence in virtual teams. When describing an influential team member, Charles, a junior team member, noted that: "his influence primarily comes through…his expertise." He also explained how his expertise garnered influence "through telling these, these stories and talking about different ways that he's done, done things elsewhere or seen things work that really…conveys that expertise and allows him to be influential." Members of virtual teams listened to people who knew about the subject area in question.

Participants' descriptions of the most influential members of their virtual teams revealed how status and influence were related. The majority of participants described their team leader when asked to discuss their most influential team members. For some participants, the manager's influence stemmed primarily from her or his role in the team and/or the organization (i.e., occupational position). For others, the key to the managers' success stemmed from personal characteristics such as the managers' energy level and their experience level. However, status characteristics comprised the majority of descriptions of influential persons.

A second way that status affected the influence process was through participants' choice of targets. Participants rarely chose their targets based on diffuse status characteristics, and if they did the choice was based on occupational position. Dorothy, a junior team member, said that she asked one team member instead of another because: "he has the expertise and the authority for…the input that I need on this particular project."

However, in the majority of cases, participants did not have a choice as to who they went to for help; either they were assigned to work with a certain individual or the target was the only person on the virtual team with the capacity to perform the task or provide the necessary information. Thus, the targets of influence attempts were often chosen based on their specific status characteristics rather than their diffuse status characteristics.

As an example of how specific status characteristics overshadowed diffuse status characteristics, Michael, a junior team member, pointed out that although the manager of his virtual team was the team member with the most influence, he was not necessarily a subject matter expert on everything. His team was comprised of five people who had different areas of expertise. In a way, this division of responsibilities (or areas of expertise) dictated whom one would go to for various needs, and therefore who people would try to influence.

Summary

Status affected the influence process in virtual teams in two ways: 1) status affected who was influential; and 2) status affected who people targeted in their influence attempts. In regards to the former, occupational position, and thus formal power, was an important source of influence in virtual teams. Other diffuse status characteristics (e.g., sex and race) played a less prominent role in determining who was influential in virtual teams. Specific status characteristics, such as communication skill, also played a role in determining who on a virtual team was influential. In regards to the latter way that status affected the influence process, virtual team members did not single out low-status team members. Instead, people were 'forced' to go to certain people for desired outcomes due to their expertise and their capacities. In this way, the status process reflected the focus within virtual teams on the task – members of virtual teams went to their expert on the topic in question to get the task done efficiently and to get it done right.

Now I will discuss additional findings of the study that are not directly related to the research questions. These findings include influence strategies used within virtual teams, participants' communication technology preferences, and the importance of culture and time in virtual teams.

Additional Findings

Due to the nature of qualitative data, the emergence of unexpected categories during data analysis is not uncommon. In the following section I will first describe participants' communication technology preferences, which relates to how they engage in their influence attempts. I will then discuss the importance of culture and time that emerged from the data, which provides a context for the added difficulty of influencing others in the virtual environment.

Communication Technology Preferences

Several participants expressed a strategic preference for communicating with their team members using certain technologies, beyond simply using certain technologies to get their targets' attention as discussed previously. This preference stemmed from a belief that certain communication technologies were more effective at conveying a message, either through their efficiency, the richness of the chosen medium, or their synchronicity. Effectively conveying messages, in addition to getting the target's attention through the use of multiple communication technologies, helped ensure that participants were not ignored and that they succeeded in their influence attempts.

Several participants indicated a preference for telephone interactions. For individuals who never met their team members FtF, speaking over the phone was a great way to make a connection with their team member and to help avoid or clear up misunderstandings. Preferences for telephone were also based on the relative richness of the media compared to other forms of virtual communication (i.e., telephone contains cues such as tone of voice) and its synchronicity. As Michael, a junior team member, stated: "I think that's probably the biggest influencer is...using the phone...over other methods of communication to sort of duplicate...or that kind of takes the place of that FtF exposure."

Other participants expressed a preference for FtF interactions whenever possible. For many, FtF interactions were critical to quickly establishing relationships with their team members. For others, including Michelle, a junior team member, FtF meetings helped ensure the success of influence attempts.

> So in this case we were able to meet person-to-person. Even though she's from out of town, she was coming to town so we made arrangements to, you know, sit down and go for lunch when she got here so we could, you know, look at one another...I can see body language, she can see body language. And that was the appropriate, I think, thing to do at that point.

Preferences for email were based in large part upon characteristics of the influence target. If the target had a habit of "forgetting" certain conversations or was especially chatty, emails provided a way to record interactions and limit the amount of time that was spent on the communication. Thus, email was seen as a way to save time, as described by Richard, a senior team member:

> What I had found that works best with him to actually get an answer on, is to lay out pros and cons of something, put it all in an email, um, because...if you get him on the phone, he can be a talker and will want to

walk through it and convince himself. And before you know it fifty minutes later you'll still be on the phone with him. You might get an answer, but you might not. So...to lay it out over email and say pros and cons, and he's normally, he's very responsive over email.

These communication technologies provided participants with a way to connect with their team members. Richer media were more effective at building relationships, while less rich media helped with efficiency. Thus communication technology preferences reflected the focus of virtual teams on both relationships and tasks. While communication technology preferences were under the control of the participants, other important aspects of their virtual teams emerged that were not under their control.

In-group – Out-group Effects

Participants described various aspects of culture and time as they related to their virtual teams. While their descriptions revealed differences amongst team members, these differences reflect in-group – out-group differences as opposed to status differences. This distinction is due to the fact that status characteristics theory only ascribes status differences if the status beliefs are accepted by the entire population (Wagner & Berger, 2002). In-group – out-group distinctions, on the other hand, reflect individuals' beliefs that their sub-group (e.g., American or European) is the high status sub-group (Tajfel, Billig, Bundy, & Flament, 1971; Tajfel, & Turner, 1986). That is, if all members of a virtual team believe that their sub-group is high status, no single sub-group can be high status because the status beliefs are not accepted by the entire group. Thus, participants' descriptions of cultural and time-related differences within their groups differ from their descriptions of status differences.

Culture

The cultural issues (e.g., availability, language differences, etc) that emerged in the current study have also been found in previous research on intercultural communication (Hall & Whyte, 2008). Several participants indicated that the impact of culture was seen primarily through the availability of their team members. These participants noted that people who worked in other countries had different holidays and that they were unavailable when they were not in the office. Language differences were also mentioned by several participants as impediments to effective collaboration. One junior team member, Michelle, mentioned impact of employing people from aboriginal cultures[4].

> But along with that are some cultural differences certainly between I guess what you call European culture versus aboriginal culture…for instance, aboriginal people who tend to never disagree. Um, and that was a hard lesson for me. But if I would ask someone a question…the polite aboriginal response is, 'Yes.' Meaning, 'Yes, I hear what you're saying.' Not necessarily, 'Yes, I agree with you'…I've learned to not ask closed-ended questions, always open-ended questions.

For the most part, however, cultural differences – like previously discussed sex and race differences – were not seen as important factors in virtual team interactions. William, a junior team member, indicated this:

> Even within our…European team we have people in England, we have, um, people in Ireland, we have, you know, people in France…there's even, you know, among the English we have a guy from northern

---

[4] Aboriginal refers to a native of northern Canada.

> England, kind of a Manchester type, and then a guy kind of, you know, of the London mold…sure there's differences in the way even people who live relatively close to each other communicate, um, in their various cultures, yeah. There's really a hodgepodge on these calls, but I would say, you know, it isn't any different than people you went to school with. Though you might be all from the same place…and you might all, if you will, share the same culture, you just have those things that are distinct about you…and you find those same things on the calls, you have the sarcastic one, and it doesn't matter where they're from, they're just kind of that person…you kind of create your own little school atmosphere, if you will, uh, oddly within these teams, even though, you know, people might try to look at this like a big cultural type difference…it's just people are people and in the end…their personality shines through much more than their culture, if you will.

Many people also felt that their organization's culture negated the impact of cultural differences. Charles, a junior team member in the U.S. military, stated: "We're all members of an organization with a very, very strong culture that I think overrides a lot of…any existing cultural differences that may have even been there before." In the next section, I will discuss the various aspects of time that emerged from the data and how they are related to culture.

Time

Time is also an important component of virtual work. While certain aspects of time were described by participants in terms of status differences (e.g., time zone differences as they affected meeting times), typically their descriptions of time reflected in-group – out-group differences. Time zone differences, deadlines, priorities, multitasking, pace, response time, time management, and timelines all played a prominent role in participants' narratives. Some aspects of time, such as those associated with time zone differences, pace, and response times, were closely related to culture. This cultural variance in the treatment of time has also been found in previous research (Hall, 1988; Hall & Whyte, 2008).

Time zone differences. Time zone differences are a significant aspect of time in virtual teams. Time zone differences were often associated with cultural differences. Participants who worked in virtual teams that spanned multiple time zones often worked with team members from different countries and different cultures. In addition, regional culture differences could arise even when team members were in the same country.

Time zone differences affected when virtual team members were able to contact one another and when people could expect responses. If the time zone difference was too great, synchronous communication might not have been possible unless team members significantly altered their workday. Dorothy, a junior team member, described the impact of time zone differences on her virtual team:

> I am working with people for example from Ireland who are five hours ahead of me. So at this moment, it's the end of their business day. Um, so I have to be very conscious of where people are in their day, and I have to be conscious of, you know, what they're waiting for from me, when I communicate with them, how I communicate with them, that type of thing.

Pace. Participants discussed pace in two different contexts. In the first, participants discussed how the pace of work varied based on the location of the virtual team members. Differences in the pace of work were attributed by participants as due to regional culture differences (e.g., New York was faster-paced than Maine). For example, Laura described how her team member differed from her in the following terms:

> He's in Maine and I'm in New York City. So, while we have the same job description, the pace of life between the two of those locations is a little different… New York is about five times as busy as Maine. So there's volume.

This finding of regional differences in pace of work is supported by previous research which identified differences in pace of life in different cities, regions, and countries (Levine, Lynch, Miyake, & Lucia, 1989; Levine & Norenzayan, 1999).

Participants also described the slower pace of working in a virtual environment. As Patricia, a senior team member, said: "This thing about not being in the same place is, it's slower communication." And, according to Charles, a junior team member, "It's just a slower process, of having to get to know people." The frustration expressed over communication delays illustrated participants' desire to accomplish their tasks.

Response time. Participants reported a concern for slow response times from people they were trying to influence. When members of virtual teams had to wait for a response, it kept them from their task. However, Richard, a senior team member, was concerned that sometimes people respond to quickly, which was associated with an entirely different set of problems: "Some people are very quick to fire back emails, off the cuff. Um, some people are very emotional, via email, and if they're pissed off they'll let…expletives fly via email and do different things like that." Richard also reported that response times were associated with cultural differences: "And response times are different…Culturally…the Mexican group, not to be stereotypical, but tends to take a little bit longer in their responses."

Deadlines. Participants described deadlines as an effective way to keep their virtual team members on task. Effective team members were those who set clear deadlines and followed up with people to ensure deadlines were being met. One senior team member, Richard, commented on the importance of deadlines in virtual teams:

> I think, um, deadlines help. That's one thing I think, more virtually than, than somewhere else. 'Cause, like I said, if you're in the same building with somebody then you can just walk down the hall and say, 'Hey, I need this.' And they say, 'When by?' And you say, 'Oh, I'm not sure.' But you can always go by the next day and say, 'Yeah, I really need this now.' Rather than when you're working virtually it's, 'Hey, I need this by close of business on Tuesday.' Or, 'I need this in an hour.' It's, I think it's very important to state urgency that sometimes is easier in person to state rather than on a virtual team.

Although deadlines are important in all work environments, there was a pervasive concern with them among the participants. Creating timelines was a specific behavior participants used in order to ensure deadlines were met. Timelines were used by participants to keep team members on task.

Multitasking. Multitasking illustrated the importance of tasks in virtual teams while also providing an example of how it could be difficult for members of virtual teams to focus on one task at a time. Participants reported instant messaging and emailing team members while participating in conference calls. For participants who multitasked on a regular basis, it was due to expectations from management. In Betty's organization, a large financial corporation based in the Southeastern United States, you had to "multitask to survive." However, participants did not view multitasking as an efficient use of time; rather it was a necessary evil.

Summary

In the previous sections I have discussed some of the unanticipated findings of the current study. These findings include strategic preferences for certain communication technologies as well as in-group – out-group differences related to culture and time. While these categories are not directly related to status and influence in virtual teams, they provide a context for virtual work.

Chapter Summary

Members of virtual teams have access to two categories of lateral influence tactics: traditional influence tactics and ambiguity reduction techniques. The most commonly used lateral influence tactics were pressure and legitimating tactics. While it was used less frequently, exchange was a successful lateral influence tactic.

The second finding was that lateral influence tactics include traditional influence tactics as well as new influence tactics that are enabled by the virtual environment and the technologies used to communicate in virtual teams. Lateral influence tactics tended to be more assertive in virtual interactions than they have been found to be in FtF interactions (i.e., the use of pressure is more common in virtual teams).

The third finding of this study is that status differences are present in the virtual environment, and these differences affect the influence process. Status affects not only who is influential in virtual teams but also who people target in their virtual influence attempts. Specific status characteristics (e.g., skills, knowledge, and expertise) were how the majority of participants differentiated between members of their virtual teams. While an awareness of diffuse status characteristics was present, they did not – with the exception of occupational position – determine who was influential or who was targeted for an influence attempt in virtual teams.

The fourth finding was that low-status virtual team members can successfully influence their team members. In fact, a higher proportion of upward influence attempts were successful compared to lateral influence attempts. The use of pressure was the most successful upward influence tactic followed by rational persuasion and legitimating tactics.

In the next chapter, I will discuss my analysis of the data. This will include a synthesis of the data into overarching themes that describe the experience of influencing virtual team members. I will also describe the influence process that emerged from the data during analysis.

DISCUSSION

In the current study, I examined influence and status in the context of virtual work teams. One purpose of this research was to identify what influence tactics are available to members of virtual teams and which are most successful. A second purpose was to understand how status affects the influence process in virtual team interactions. This study will provide a better understanding of the influence process in virtual teams and will help develop guidelines for successful virtual teamwork. The study was based on the following four research questions.

1. What influence tactics are available to individuals who interact in virtual teams?
2. How are these similar or different to those available to individuals who interact with FtF?
3. How does status affect the influence process in virtual teams?
4. How do low-status individuals successfully exert power over other members of their virtual teams?

In this chapter I will interpret the findings and tie the results back to previous research. While power was a focus throughout the course of this study, it was never explicitly addressed by participants, primarily because power is a topic with which most people are uncomfortable. It is also difficult for people to articulate issues of power. For example, when Kimberly was asked if she had a plan or strategy to get what she wanted from a coworker, she responded: "Did I have a plan?...I wouldn't say, because of the way you're phrasing it...Marla, makes it, it sounds devious and I don't think you mean it that way." However, the pervasiveness of power issues (in a prescriptive sense) in virtual teams were illustrated by participants in their descriptions of the influence tactics they used (primarily pressure and legitimating tactics) and the status characteristics that distinguished team members from one another (e.g., occupational position).

The chapter is organized by the following analytic categories. First, there is a tendency to use more assertive influence tactics in virtual teams than has been found in research on face-to-face interactions. Second, the success rate of influence tactics varies by the direction of the influence attempt. Third, specific status characteristics are more relevant for members of virtual teams than diffuse status characteristics. Fourth, there is both a relationship orientation and a task orientation in virtual teams which creates a tension. And finally, I will present an initial influence process for virtual teams. The discussion will refer to the literature on status, influence tactics, CMC, and virtual teams. The implications of the results are intended to illustrate the opportunities that virtual work creates for individuals. The chapter will conclude with a summary of the interpretation and analysis.

Tendency to Use Assertive Influence Tactics

The three most frequently used influence tactics in virtual teams are pressure, legitimating tactics, and rational persuasion. Pressure and legitimating tactics are more assertive than the influence tactics most frequently used in face-to-face influence attempts (i.e., consultation, rational persuasion, and inspirational appeals) (Yukl & Falbe, 1990). In addition, pressure is not as successful when enacted face-to-face (Yukl & Falbe, 1990).

One reason for these assertive tactics may be the more ambiguous nature of the virtual environment compared to the face-to-face environment; it makes it easier for targets to ignore influence attempts. Subsequently, influencers may need to use more pressure. This finding nonetheless runs counter to previous research of influence in virtual teams.

The finding that members of virtual teams frequently, and successfully, use assertive influence tactics conflicts with the conclusions drawn by Elron and Vigoda-Gadot (2006), who claimed that the use of hard influence tactics is milder in virtual teams. Like the current study, Elron and Vigoda-Gadot's (2006) results were based on interviews of virtual team members. Participants in their study were drawn from eight teams in two organizations, with two to three members from each team; whereas the current study did not interview members of the same team. One explanation for the differences in results is a potential bias in responses in which respondents understate their use of socially undesirable influence tactics, such as pressure, a possibility when researching influence that was noted by Yukl and Falbe (1990). Participants in Elron and Vigoda-Gadot's (2006) study may have minimized their discussion of behavior considered to be unacceptable.

While social desirability is also a possibility in the current study, several factors about the current study indicate that this is not the case. First, participants were asked to report on specific behaviors that had occurred recently in order to get a more accurate response. Second, participants in the current study were vocal about their reticence to talk about influence and power, but they nonetheless provided valuable information and insight regarding the influence process in their virtual teams. Third, participants reported using less socially desirable tactics more than socially desirable ones which indicates that they were not as concerned with responding in a socially desirable manner.

A second explanation for the discrepancy between the results of the current study and that of Elron and Vigoda-Gadot (2006) comes from social information processing theory which suggests that power relations may develop over time (c.f., Walther, 1995). If Elron and Vigoda-Gadot had studied members of well-established virtual teams, as was the case in the current study, influence attempts may have been harder and more prevalent. Once a team history is established and members are more comfortable communicating with one another, power plays and the use of influence tactics may become more common. This is supported in the current study by the emergence of the relationship building influence strategy.

A third explanation for the discrepancy between the results of the current study and that of Elron and Vigoda-Gadot (2006) is the impact of culture. In their study, the majority of participants were Israelis based in Israel, while in the current study the majority of participants were North Americans based in North America. Support for this explanation comes from Hofstede's (1983) dimensions of national cultures in which assertiveness is a connotation of the masculinity-femininity dimension. Hofstede (1983) showed that the United States and Israel differ in their level of masculinity, which would affect the likelihood of using assertive influence tactics.

Participants' descriptions of the inherent ambiguity of the virtual environment, along with the increased ease of ignoring influence attempts, provide an explanation as to the increased reliance on various forms of pressure in virtual teams. These descriptions also provide explanations as to how the current results differ from those of previous research. Below I will discuss explanations as to why the success rate of influence tactics varied by the direction of the influence attempt.

Variation in Success Rate of Influence Tactics

Participants' accounts of their influence attempts revealed different success rates depending on the direction of the influence attempt. Successful influence attempts reflected the power of the actor, in a prescriptive sense. The fact that lateral tactics were only effective about half of the time, and that upward and downward tactics had a much higher rate of success, indicates that there is some pushback from team members when the influence attempt comes from a peer. In the following paragraphs, I will explore why this may be the case.

One explanation as to why approximately half of the lateral influence attempts were successful could be related to the relationship building influence strategy. Relationships with peers have become increasingly important due to the increased necessity of collaboration (Forret & Love, 2008). Although personal relationships are more difficult to establish in virtual teams than in face-to-face teams, they do develop over time (Walther, 1995). It is more difficult for these relationships to develop in a virtual context if trust is not established (Long, Kohut, & Picherit-Duthler, 2005; Long, Picherit-Duthler, & Duthler, 2009).

Personal relationships provide peers with a reason to comply with an influence attempt (Yukl et al, 1995): what people may not do for a coworker, they will do for a friend. In support of this explanation, all of the lateral influence attempts that involved building relationships were successful. Kimberly discussed the importance of building relationships with her peers: "Again, a lot of initial time invested in getting, you know, getting to know each other, building up trust…and then looking at…what resources we have that we can share. So…that enables us to use each other as peer coaches." The possibility that the success of lateral influence tactics is heavily reliant on personal relationships and that these relationships are harder to establish provide explanations as to why lateral influence tactics, in general, were less successful and why those that incorporated relationship building were successful.

However, the difficulty and necessity of establishing peer relationships alone does not account for the varying success of influence tactics. The majority of downward and upward influence attempts were successful most likely for structural reasons. Supervisors have formal power and the ability to reward and punish their subordinates, thereby increasing the probability that targets will comply with their requests.

While subordinates do not have formal power, the role of supervisors obliges them to help their subordinates. In most instances described by the participants, the objective of upward influence tactics is to seek information or resources necessary to do a task. It is part of the supervisor's role to comply with these types of requests.

Therefore, a second explanation why lateral influence tactics had a lower rate of success could be that peers are not structurally required to help one another. This lack of structure is also present in face-to-face teams, but it is exacerbated by the virtual environment. In addition, complying with a request from a peer involves the target giving that coworker influence over her or himself. Peers may be reluctant to comply with requests because it involves giving power and influence to people who are supposed to be their status equals.

In sum, while establishing relationships was important in all three directions, it appears to be most important in lateral influence attempts. This could be due to an absence of structural reasons to help peers, which existed in the upward and downward directions. Structural reasons for complying with influence requests provide an explanation for the high rate of success of upward and downward influence tactics, while the establishment of relationships accounts for the success of lateral influence tactics.

Increased Relevance of Specific Status Characteristics

Status differences in the current study were based on prescriptive power differentials (e.g., occupational position or expertise). To explore the sensitive topic of status, participants were asked what differentiated members of their virtual teams from one another and to provide descriptions of their influence targets and their most influential team members. In response, participants indicated that members of virtual teams are differentiated from one another based on several characteristics which can be classified as diffuse and specific status characteristics, as well as status cues. These reported status differences affect the influence process in the following ways: 1) status affects who is influential; and 2) status affects who people target in their influence attempts. Below, I will provide explanations as to why specific status characteristics were more relevant for members of virtual teams than were diffuse status characteristics. To do this, I will begin with explanations as to why diffuse status characteristics may have been less important in the virtual environment.

One aspect of status in virtual teams that differentiated it from status in face-to-face interactions was that diffuse status characteristics were not as important in determining an influence target as were specific status characteristics. One explanation as to why diffuse status characteristics are not as relevant in the virtual environment comes from Driskell and colleagues' (2003) input-process-output model. As part of this process, they propose three mechanisms through which technological mediation may impact status processes: 1) CMC may *block* the transmission of status characteristics and status cues; 2) the effects of status characteristics and status cues may be *dampened* in virtual environments; and 3) status expectations may not be *translated* into behavior due to weakened norms. In the current study, certain status cues and diffuse status characteristics were not available (blocked), and others such as gender and race were dampened or simply not translated.

A second explanation as to the diminished importance of diffuse status characteristics is the limited availability of identity cues in the virtual environment. SIP asserts that it takes longer for identity cues to manifest via computer-mediated communication and for relationships to develop (Walther, 1995). These identity cues are closely related to diffuse status characteristics in that they are identifying features of virtual team members that are not necessarily immediately apparent.

While diffuse status characteristics appeared to be less relevant in virtual teams, it could also be that specific status characteristics were much more important in this context. In support of this explanation, informational advantage was seen by participants as another source of power (cf. Baldwin et al, 2009). Informational advantage is closely related to personal power (i.e., it is based upon expertise). This underscores the important role personal power plays in interpersonal interactions compared to formal power, which is based upon occupational position, a diffuse status characteristic.

Another explanation as to the increased importance of specific status characteristics comes from status characteristics theory. According to this theory, power and prestige structures emerge based on the status characteristics that are activated small group interactions (Wagner & Berger, 2002). Upon what are these structures based (i.e., how is influence distributed) in virtual teams when a variety of identity cues are not available? Diffuse status characteristics were found to be less relevant in virtual team interactions than they typically are in face-to-face interactions. The distribution of influence in virtual teams was instead based upon information that is readily available in virtual teams: the competence of the team members.

The relevance of status characteristics in general, and the importance of specific status characteristics in particular, also could have stemmed from the task-focused nature of virtual teams. For example, age and occupational position were the diffuse status characteristics most likely to be activated in the virtual environment. This is most likely because they related to experience and the person's capabilities. Specific status characteristics, such as expertise, knowledge, and skills, also reflect the task orientation of virtual team members. In addition, tech savviness reflects a characteristic that is uniquely relevant in the virtual environment when trying to accomplish a task using communication technologies.

In sum, specific status characteristics were more prevalent in participants' descriptions of their influence targets than were diffuse status characteristics. This finding may be due to the fact that diffuse status characteristics are less available in the virtual environment. It may also be due to the fact that specific status characteristics are more important in the virtual environment given the available cues and the task-focused nature of virtual teams. Below I will discuss this task orientation and its tension with the relationship orientation of virtual team members.

A Dual Orientation

A major theme that emerged in the study was tension between the relationship orientation and task orientation of virtual team members. Task orientation is a focus on the virtual team's purpose or goal as opposed to the social elements (i.e., a relationship orientation) of virtual team members' interaction. Thus, having a relationship orientation and a task relationship creates a sort of tension which members of virtual teams must balance. In fact, Dube and Robey (2008) identified this tension as one of the five paradoxes of virtual teamwork (i.e., "task-oriented virtual teamwork succeeds through social interactions").

In their descriptions of communication technology preferences, participants' explanations for their preferences highlighted the tension between relationship orientation and task orientation. Richer media were described as more effective at building relationships, while less rich media were described as helping with efficiency because they would not be stuck in lengthy conversations. This preference for leaner (or less rich) media runs counter to Media Richness Theory given the equivocal nature of the virtual environment (Daft & Lengel, 1986). In addition, it runs counter to Media Choice Theory which argues that people prefer to interact with others on a more person level, which involves a preference for face-to-face communication (Mullen, 2005; Murray & Peyrefitte, 2007).

However, in support of Media Choice Theory, many participants reported that if they had had the opportunity to interact face-to-face during their lateral influence attempts, they would have taken it. Participants believed face-to-face interaction reduced the number of communication issues and enabled the influence process to be accomplished more quickly. Interestingly, they didn't feel that face-to-face interaction would have affected how they approached the target or how the target would have responded. This may be related to the fact that participants most often reported using tactics they would have used face-to-face interactions. Richer media could have made these virtual influence attempts very similar to face-to-face influence attempts. In the following paragraphs I will discuss what the importance of relationship and task orientations means for virtual team members.

Relationship Orientation

Relationship and task orientations are not necessarily unique to virtual teams; however, the nature of virtual teams is such that extra care has to be taken when building relationships. It takes longer to establish and build relationships using computer-mediated communication (Walther, 1995). The extra effort required to establish relationships provides an explanation as to why relationship building strategies were so prominent in the interviews.

Why would virtual team members expend so much effort on building relationships? The current study, in support of previous research, provides answers to this question. First, relationship building enables individuals to establish personal power (Powell et al, 2004) and, therefore, be more successful when using influence tactics. Kimberly explained how building relationships and establishing trust enhanced her ability to influence her team members:

> So how do I get others to do what they need to do? Um, persistence…a key factor I found is developing a relationship early on prevents problems later on…We have found that investing time with our colleagues to

develop, um, awareness of the individuals' strengths, their perspectives, their particular, um, challenges, and really cultivating trust pretty much alleviates, uh, barriers farther down the line.

A second reason that virtual team members expend effort establishing and maintaining relationships with their coworkers is that a relationship orientation can ultimately help virtual teams reach their task goals through increased efficiency. Indeed, in Pauleen's (2003) study, virtual team leaders reported that establishing these relationships was necessary before work on the task could even be commenced.

Task Orientation

The reason for the task orientation of virtual team members is much easier to explain. The task is the purpose of the virtual team; it is the reason why the virtual team members are interacting with one another (Jarvenpaa & Shaw, 1998). The task orientation of virtual teams was made evident through their descriptions of their virtual team members. There was a focus by participants on specific, task-relevant, status characteristics, such as effective communication skills, rather than demographic information such as gender and race, or personal information such as hobbies. The focus on specific status characteristics reflected the task-oriented nature of virtual teams.

Participants' descriptions of the various components of time also reflect the task-focused nature of virtual teams. Time zone differences, deadlines, priorities, multitasking, pace, response time, time management, and timelines all played a prominent role in participants' narratives. These different components of time reflect a task-focus because they affected how efficiently tasks were accomplished.

Time zone differences, in particular, reflected the theme of task orientation because participants focused on how these differences either expedited work on the task or slowed down progress on the task. If work was handed off efficiently, then there was always someone available to work on the task (i.e., When Person A left work for the day, she sent her work to Person B who was just starting his day, and vice versa). Time zone differences could also impede the progress of work if this handoff was not made properly. In addition, it was difficult to provide effective feedback to teammates who were only working when the person giving the feedback was "off the clock" because it could not be given in real time.

As mentioned in the results, the task-focused nature of virtual teams provides possible explanations as to how virtual team members choose their targets and how low-status team members can be so successful in their influence attempts. Rather than choosing a low-status – or "easy" – target, participants reported targeting team members who could accomplish the task. This also explains how low-status team members garnered more influence – the task-focused nature of the team may have overshadowed the status differences.

Task orientation has been found in other research on power in virtual teams. In Elron and Vigoda-Gadot's (2006) study of politics and influence in virtual teams, task orientation emerged as a mediator in the relationship between characteristics of virtual teams and politics and influence tactics. These researchers noted that one of the outcomes of a task focus is that there is less time for informal socializing, which in turn can result in higher effectiveness.

Summary

As can be seen from the current study, effective virtual teams require more than task orientation; they also need an explicit relationship orientation. These two needs in virtual teams conflict with one another, creating a sort of tension. Virtual teams must decide how to best spend their time: focusing on the task or on relationships. Given the time-consuming nature of relationship building using computer-mediated communication and the pressures to complete tasks (e.g., strictly enforced deadlines), there is little extrinsic motivation to spend time establishing and maintaining relationships. Successful influencers have realized the importance of relationships to the effective completion of their assigned tasks.

The Influence Process in Virtual Teams

Finally, I present a model for the influence process in virtual teams (See Figure 1). This process reflects the prescriptive nature of power (French & Raven, 1959), as opposed to its more insidious form. Nonetheless, this model will help leaders and members of virtual teams understand how influence works in this relatively new environment. It will also help practitioners identify best practices in virtual teams (e.g., relationship building). The model contributes to the literature by highlighting the various ways in which technology affects the influence process in virtual teams (e.g., through the strategic use of technology to get the target's attention and communication technology preference). It also introduces a new category of influence tactic: ambiguity reduction techniques.

The first step of the influence process in virtual teams revolves around the strategic use of technology to get the target's attention. The second step of the influence process revolves around the strategy of building relationships with team members.

Members get to know each other and build trust. The third step of the influence process revolves around the influence tactics themselves. The agent must choose amongst previously identified face-to-face influence tactics and ambiguity reduction techniques. The influence process is affected by the agent's communication technology preferences and status. The influence process will be described in more detail below.

The first step of the influence process is to get the target's attention through the strategic use of technology. As was stated in the results, it is more difficult for the agent to get the target's attention and much easier for the target to ignore the influence attempt when the interaction occurs in the virtual environment. Thus, getting the target's attention emerged as part of the influence process in virtual teams even though it is not an explicit part of the influence process in face-to-face teams. Indeed, in a face-to-face team, one merely needs to be present in order to get the target's attention.

Influence strategies (i.e., relationship building and documenting communications) are important next steps in the influence process. In order to build relationships, agents (i.e., those who are influencing) must first get to know their team members. Second, influencers must establish trust with their team members. The third and final phase of the building relationships process is ongoing – influencers must work to maintain the relationships with their team members in order to successfully influence them.

While building relationships is not a requisite step in the influence process per se, it does help to ensure the success of the influence attempt. In the current study, influence attempts were successful each time a relationship building strategy was enacted. However, only 17 of the 23 participants mentioned relationship building in some form, and it was only mentioned in direct connection with 16 of the influence attempts described by participants (as opposed to discussing the importance of relationship building in general) This indicates that relationship building is not a universal influence strategy.

Once relationships have been established, it is easier for agents to successfully use influence tactics. One reason for this success is that personal appeals and exchange tactics are more easily used. Building relationships enables the use of personal appeals due to the fact that the use this tactic involves appealing to a person's feelings of loyalty and friendship (Yukl et al, 1995). Building relationships also enables the use of exchange through the establishment of trust. In order to effectively use exchange, the target must trust that the agent will reciprocate the favor at a later time or will share the benefits that result from helping the agent (Yukl et al, 1995). A second reason that influence tactics are more successful once relationships with targets have been established is that targets are more likely to comply with requests from their friends.

Building relationships takes additional time when using computer-mediated communication. What happens if a relationship has not yet been established or if the agent does not use this strategy as part of her or his influence process? How can agents be successful if a relationship has not yet been established? One explanation is that assertive techniques are more acceptable in the virtual environment. In this context they are used primarily to get the target's attention rather than to coerce or bully the target.

Documenting communications is also an influence strategy that agents may choose to enact before an influence attempt. Keeping an email trail allows agents to have a written record of the work they have done and the work they have requested from others. This strategy is something that some participants reported using on a consistent basis; however, it was not mentioned by all participants.

Regardless of whether or not a relationship has been established with the target or communications with the target have been documented, using an influence tactic is a requisite step in the influence process of virtual team members. These team members have access to influence tactics that were identified in previous face-to-face research (cf., Kipnis et al, 1980; Yukl & Falbe, 1990); however, many of these tactics have been modified in order to be used in the virtual environment. In addition, the ambiguous nature of the virtual environment has prompted the use of ambiguity reduction techniques. If the first influence tactic is unsuccessful, virtual team members either switch to a different tactic or follow up with the target.

Throughout the influence process, characteristics of the agent and target affect the various components of the process. These variables include the communication technology preferences of the agents and status. Below I will discuss the roles of these variables in greater detail.

Communication technology preference affects both the influence strategy and influence tactics components of the influence process in virtual teams. This is due to the fact that members of virtual teams interact for the most part, if not entirely, through communication technologies. At each stage of the influence process, the agent must choose which communication technology she or he will use. As maintained by the cognitive model of media choice theory, participants generally preferred richer media because it is high in social presence (i.e., communication is more likely to be synchronous and there are more identity cues available than in leaner media) (Robert &

Dennis, 2005). Media choice affects how receptive a target will be to the influence attempt (i.e., whether the agent can establish a relationship or successfully use an influence tactic). However, not all participants preferred rich media. Some participants preferred lean media such as email because of its efficiency. These different preferences reflect the tension between relationship orientation and task orientation.

The effects of status, on the other hand, are primarily seen in the influence tactics stage of the influence process in virtual teams. As seen in the results of the current study, status is one of the many factors (i.e., communication technology choice, whether a relationship has been established, and choice of influence tactic) that affects the success of the influence attempt. Regardless of the status relationship between the agent and the target (e.g., upward, lateral, or downward), relationships can be established.

The influence process described above ties the results of the current study to past theory and research. The model also serves as an indicator of the practical implications from the results of this study. In the next section, I will more explicitly state what the practical implications of this study are for both low-status and high-status members of virtual teams.

Practical Implications

The practical implications of this study are focused on the ways in which members of virtual teams – low-status members in particular – can improve their effectiveness and successfully influence their team members. It is important for all members of a virtual team to be heard to increase the quality of decision making (cf., DeSanctis & Gallupe, 1987), but it is especially difficult for low-status individuals to be heard. I suggest that members of virtual teams, particularly low-status team members, use more assertive influence tactics as a way to be heard by their target and increase the likelihood of influence success. Another strategy for successful influence is relationship building.

A second practical implication focuses on how virtual team leaders can increase relationship building in their teams. Several participants mentioned that building relationships takes longer in virtual teams than in face-to-face teams. One way that virtual team leaders can support relationship building in their teams is to create the opportunity for the team to come together face-to-face at least once a year in order to establish and maintain relationships amongst team members (Dube & Robey, 2008). Another step that virtual team leaders can take to maintain the relationships of their virtual team members is to have team building exercises at the beginning of meetings, as did Karen, in order to allow for social communication before focus on the task commences. Fostering social relationships is critical for the realization of intangible benefits in virtual teams, such as a committed and talented workforce (Long et al, 2005; Long et al, 2009). Virtual team socialization will also help foster relationships and enable effective communication (Long et al, 2005; Long et al, 2009).

Limitations and Future Research

Although the qualitative nature of the current study produced valuable findings regarding the influence process in virtual teams, there are also some limitations to the research. The first limitation pertains to participant selection and characteristics of the participants. Data collection was limited to 23 members of virtual teams who self-selected into the study. Self-selection poses a potential problem because people who agree to participate tend to be those with positive views of their virtual teams and virtual team members (Rogelberg, Luong, Sederberg, & Cristol, 2000). Based on multiple comments by participants, who indicated frustrations with their virtual team members, this does not seem to be a concern. Therefore, it does not appear that the participants in the current study have an overly positive view of virtual work and their virtual teams.

Additionally, these participants were all white and were all members of Western cultures. Participants did belong to virtual teams that consisted of members of other races, ethnicities, and cultural backgrounds; however, given the focus of the current study on status differences, future research would benefit from a more diverse group of participants. Interviewing participants with more diverse backgrounds will lend further support to the current influence process model.

A third limitation of the current study is that all data was self-reported. Self-report data is subject to social desirability effects, which are even more of a concern given the sensitive nature of influence as a research topic. However, in order to prevent social desirability, participants were asked to report on specific behaviors that had occurred recently. Future research should use both agent and target accounts, as did Yukl and Falbe (1990).

A fourth limitation of the study is that the effects of many traditional status differences (e.g., gender and race) are implicit. Although participants reported that these characteristics of their virtual team members did not impact their interactions, any effects may occur at the subconscious or unconscious level. Future research should observe members of virtual teams influencing one another in order to see the implicit effects of status as opposed to relying on reports from participants.

Future research should also further examine the model of the influence process. Structural equation modeling can verify the causal relationships proposed in the current model. Quantitative research will also help verify the role of communication technology preference and status in the model.

Conclusion

This research project has allowed me to interact with members of a variety of different virtual teams and organizations. One of the main takeaways from these conversations has been that communication technology and the virtual environment are changing how we work with one another in a variety of ways. How we influence one another in virtual teams is no exception to this transformation.

Needless to say these technologies are not being adopted at the same rate in all industries or organizations; however, slowly but surely technology is transforming how the entire workforce operates. Susan illustrates how this transformation has occurred in her organization:

> I think from a technology perspective…our company has been, it's an old, you know, industrial company. And we're just recently repaved our way. It took like ten years just to transform our company into something a little bit different where we're not heavy machinery anymore. We're more of a…diversified portfolio. And with that comes new ways to…be interconnected with everyone.

Changes in technology have impacted more than just how we run our businesses and communicate. They have also impacted the expectations we have when working with one another. Technology provides us with the ability to communicate with geographically dispersed coworkers and receive an immediate response. We are often able to remain in constant contact with one another. In some organizations, this ability to receive immediate responses has created expectations for them. This expectation can be seen in several participants' descriptions of multitasking requirements in their organizations. Responding to email inquiries while participating in a conference call is now a common occurrence in many workplaces.

The immediacy of virtual communication has also changed people's communication preferences. As demonstrated by Richard and other participants, people have begun to see email and other written forms of communication as more efficient than FtF or verbal communication. These preferences raise an interesting challenge to Media Richness Theory (Daft & Lengel, 1986; Mullen, 2005) and illustrate one way in which workers have begun to adapt to this new working environment.

Workers have adapted to the virtual environment in other ways, as was shown in the current study. Figure 1 illustrates differences in the virtual influence process. First, people must now work harder to get their coworkers' attention in order to even have a chance to influence them. Second, personal relationships become much more important in the virtual influence process, even though they take more time to establish in this context. Third, the virtual environment has allowed workers to adapt old ways of influencing others and to also create new influence tactics. The ambiguous nature of the virtual environment necessitates the use of more assertive influence tactics.

Status Characteristics Theory is still relevant to research on virtual teams, although the relevant status characteristics differ from what is expected in this new environment. Status Characteristics Theory states that diffuse status characteristics form expectations for the immediate interaction when no other information is presen, and once more information is present, other, more situation-relevant characteristics determine status (Wagner & Berger, 2002).. SIP suggests that it will take longer for this additional information (i.e., identity cues) to emerge in the virtual environment (Walther, 1995). Based on these theories, one would expect that expectations for virtual team members would be based more on diffuse, rather than specific, status characteristics. However, this was not the case. As stated previously, specific status characteristics determined the status distribution in virtual teams. Thus, virtual teams are changing the way we view and apply Status Characteristics Theory.

How can specific status characteristics be activated in an environment with so few identity cues? I believe this is due to swift trust, a concept which argues that relationships can quickly be established in the virtual environment – even more quickly and intimately than in face-to-face interactions (Jarvenpaa, Knoll, & Leidner, 1998). The quick establishment of these relationships provides team members with the information necessary to form expectations of their teammates based on task relevant characteristics rather than race, gender, or other diffuse status characteristics

The status relationship between actor and target now has more of an impact on the likelihood of a successful influence attempt. Influencing peers has become more difficult than influencing superiors. In addition, high status now stems primarily from the possession of job-relevant knowledge and skills. This finding aligns with the knowledge-based economy into which we have relatively recently transitioned. However, given the implicit nature of diffuse status characteristics, additional research is needed to see what role these characteristics play in virtual teams.

In the literature review, I gave two possible explanations for the relationship between status and influence tactics. One possibility was that influence tactics could be constrained by status characteristics. In other words, the success of an influence tactic may vary based upon the status of the person employing it. However, pressure and rational persuasion proved most effective for each influence direction, suggesting this explanation is not valid.

Another explanation I gave was that status is a type of influence tactic and that those who have higher status exert greater influence. My results do not indicate that this is the case. First, status is not an influence tactic. Instead, as is shown in Figure 1, status impacts the choice of influence tactic. Second, while high status individuals were very successful, low status team members were also successful a majority of the time. The relationship between status and influence, then, is that one's status affects one's choice of influence tactic.

In sum, influence is an important and pervasive component of interactions, including those interactions that occur virtually. Those who are influential are able to get what they want from others. Without a certain degree of influence, participants in virtual teams may not be able to work as effectively. I conclude that building relationships, being assertive, and effectively using technology can allow members of virtual teams to be influential.

REFERENCES

Anderson, C., & Kilduff, G.J. (2009). Why do dominant personalities attain influence in FtF groups? The competence-signaling effects of trait dominance. *Journal of Personality and Social Psychology, 96,* 491-503.

Anderson, C., Spataro, S.E., & Flynn, F.J. (2008). Personality and organizational culture as determinants of influence. *Journal of Applied Psychology, 93,* 702-710.

Avolio, B.J., & Kahai, S.S. (2003). Adding the "E" to e-leadership: How it may impact your leadership. *Organizational Dynamics, 31,* 325-338.

Avolio, B. J., Kahai, S.S., & Dodge, G. E. (2000). E-Leadership: Implications for theory, research, and practice. *Leadership Quarterly, 11,* 615.

Baldwin, A.S., Kiviniemi, M.T., & Snyder, M. (2009). A subtle source of power: The effect of having an expectation on anticipated personal power. *The Journal of Social Psychology, 149,* 82-104.

Bass, B.M. (1960). *Leadership, psychology, and organizational behavior.* New York: Harper & Row.

Bell, B. S., & Kozlowski, S. W. J. (2002). A typology of virtual teams: Implications for effective leadership. *Group & Organization Management, 27,* 14.

Berger, J., Cohen, B.P., & Zelditch, M., Jr. (1972). Status characteristics and social interaction. *American Sociological Review, 37,* 241-255.

Berger, J., Ridgeway, C.L., Fisek, M.H., & Norman, R.Z. (1998). The legitimation and delegitimation of power and prestige orders. *American Sociological Review, 63,* 379-405.

Bonito, J.A., DeCamp, M.H., & Ruppel, E.K. (2008). The process of information sharing in small groups: Application of a local model. *Communication Monographs, 75,* 136-157.

Cascio, W.F., & Shurygailo, S. (2003). E-leadership and virtual teams. *Organizational Dynamics, 31,* 362-376.

Charmaz, K. (2006). *Constructing grounded theory: A practical guide through qualitative analysis.* London: Sage.

Chidambaram, L. (1996). Relational development in computer-supported groups. *MIS Quarterly, 20,* 143-163.

Chidambaram, L., & Bostrom, R. (1993). Evolution of group performance over time: A repeated measures study of GDSS effects. *Journal of Organization Computing, 3*, 443-469.

Clark, H.H., & Brennan, S.E. (1991). Grounding in communication. In L.B. Resnick, J.M. Levine, & S.D. Teasley (Eds.), *Perspectives on socially shared cognition* (pp. 127-149). Washington, DC: American Psychological Association.

Connaughton, S. L., & Shuffler, M. (2007). Multinational and multicultural distributed teams: A review and future agenda. *Small Group Research, 38*, 387-412.

Constant, D., Sproull, L., & Kiesler, S. (1996). The kindness of strangers: The usefulness of electronic weak ties for technical advice. *Organization Science, 7*, 119-135.

Corbin, J., & Strauss, A. (1990). Grounded theory research: Procedures, canons, and evaluative criteria. *Qualitative Sociology, 13*, 3-21.

Cramton, C.D. (2001). The mutual knowledge problem and its consequences in geographically dispersed teams. *Organization Science, 12*, 346-371.

Culnan, M.J., & Markus, M.L. (1987). Information technologies. In F.M. Jablin, L.L. Putnam, K.H. Roberts, & L.W. Porter (Eds.), *Handbook of organizational communication: An interdisciplinary perspective* (pp. 420-443). Newbury Park, CA: Sage.

Daft, R.L., & Lengel, R.H. (1986). Organizational information requirements, media richness and structural design. *Management Science, 32*, 554-571.

deCharms, R., & Muir, M.S. (1978). Motivation: Social approaches. *Annual Review of Psychology, 29*, 91-113.

DeSanctis, G., & Gallupe, R.B. (1987). A foundation for the study of group decision support systems. *Management Science, 33*, 589-609.

Driskell, J.E., Radtke, P.H., & Salas, E. (2003). Virtual teams: Effects of technological mediation on team performance. *Group Dynamics: Theory, Research, and Practice, 7*, 297-323.

Dube, L, & Robey, D. (2008). Surviving the paradoxes of virtual teamwork. *Information Systems Journal, 19*, 3-30.

Dubrovsky, V.H., Kiesler, S., & Sethna, B.N. (1991). The equalization phenomenon: Status effects in computer-mediated and FtF decision-making groups. *Human-Computer Interaction, 6*, 119-146.

Elron, E. & Vigoda-Gadot, E. (2006). Influence and political processes in cyberspace: The case of global virtual teams. *International Journal of Cross Cultural Management, 6,* 295-317.

Forret, M., & Love, M.S. (2008). Employee justice perceptions and coworker relationships. *Leadership & Organization Development Journal, 29,* 248-260.

French, J.R.P., Jr., & Raven, B. (1959). The bases of social power. In D. Cartwright & A. Zander (Eds.), *Group dynamics: Research and theory* (pp. 607-623). New York: Harper & Row.

Hall, E.T. (1988). The hidden dimensions of time and space in today's world. In F. Poyatos (Ed.), *Cross-cultural perspectives in nonverbal communication* (pp. 145-152). Ashland, OH: Hogrefe & Huber Publishers.

Hall, E.T., & Whyte, W.F. (2008). Intercultural communication. In C.D. Mortensen (Ed.), *Communication theory* (2nd ed., pp. 403-419). Piscataway, NJ: Transaction Publishers.

Hambley, L.A., O'Neill, T.A., Kline, T.J.B. (2007). Virtual team leadership: The effects of leadership style and communication medium on team interaction styles and outcomes. *Organizational Behavior and Human Decision Processes, 103,* 1-20.

Hayne, S.C., Pollard, C.E., & Rice, R.E. (2003). Identification of comment authorship in anonymous group support systems. *Journal of Management Information Systems, 20,* 301-329.

Hertel, G., Geister, S., & Kondradt, U. (2005). Managing virtual teams: A review of current empirical literature. *Human Resource Management Review, 15,* 69-95.

Hinds, P.J., & Bailey, D.E. (2003). Out of sight, out of sync: Understanding conflict in distributed teams. *Organization Science, 14,* 615-632.

Hinds, P.J., & Kiesler, S. (1995). Communication across boundaries: Work, structure, and use of communication technologies in a large organization. *Organization Science, 6,* 373-393.

Hinds, P.J., & Mortensen, M. (2005). Understanding conflict in geographically distributed teams: The moderating effects of shared identity, shared context, and spontaneous communication. *Organization Science, 16,* 290-307.

Hofstede, G. (1983). National cultures in four dimensions: A research-based theory of cultural differences among nations.

Hollingshead, A.B. (1996). Information suppression and status persistence in group decision making. *Human Communication Research, 23,* 193-219.

Jackson, K.M., & Trochim, W.M.K. (2002). Concept mapping as an alternative approach for the analysis of open-ended survey responses. *Organizational Research Methods, 5,* 307-336.

Janis, I.L., & Mann, L. (1977). *Decision making: A psychological analysis of conflict, choice, and commitment.* New York: Free Press.

Jarvenpaa, S.L., & Leidner, D.E. (1999). Communication and trust in global virtual teams. *Organization Science, 10,* 791-815.

Jarvenpaa, S.L., & Shaw, T.R. (1998). Global virtual teams: Integrating models of trust. In P. Sieber & J. Griese (Eds.), *Organizational Virtualness* (pp.35-52). Bern, Switzerland: Simowa Verlag Bern.

Jarvenpaa, S.L., Knoll, K., & Leidner, D.E. (1998). Is anybody out there? Antecedents of trust in global virtual teams. *Journal of Management Information Systems, 14,* 29-64.

Kayworth, T. R., & Leidner, D. E. (2001-2002). Leadership effectiveness in global virtual teams. *Journal of Management Information Systems, 18,* 7-40.

Kayworth, T., & Leidner, D. (2000). The global virtual manager: A prescription for success. *European Management Journal, 18, 183-194.*

Kemper, T.D., & Collins, R. (1990). Dimensions of microinteraction. *The American Journal of Sociology, 96,* 32-68.

Kiesler, S., Siegel, J., & McGuire, T.W. (1984). Social psychological aspects of computer-mediated communication. *American Psychologist, 39,* 1123-1134.

Kipnis, D., Schmidt, S.M., & Wilkinson, I. (1980). Intraorganizational influence tactics: Explorations in getting one's way. *Journal of Applied Psychology, 65,* 440-452.

Kirkman, B.L., Rosen, B., Tesluk, P.E., & Gibson, C.B. (2004). The impact of team empowerment on virtual team performance: The moderating role of FtF interaction. *Academy of Management Journal, 47,* 175-192.

Lazarsfeld, P. F. (1944). The controversy over detailed interviews. *Public Opinion Quarterly, 8,* 38-60.

Levine, R. V., Lynch, K., Miyake, K., & Lucia, M. (1989). The Type A city: Coronary heart disease and the pace of life. *Journal of Behavioral Medicine, 12,* 509-524.

Levine, R.V., & Norenzayan, A. (1999). The Pace of life in 31 countries. *Journal of Cross Cultural Psychology, 30,* 178-205.

Lind, M.R. (1999). The gender impact of temporary virtual work groups. *IEE Transactions on Professional Communication, 42*, 276-285.

Lindlof, T.R., & Taylor, B.C. (2002). *Qualitative communication research methods* (2nd ed.). Thousand Oaks, CA: Sage.

Long, S.D., Kohut, G.F., & Picherit-Duthler, G. (2005). Newcomer assimilation in virtual team socialization. In S.H. Godar & S. Pixy Ferris (Eds.), Virtual and collaborative teams (pp. 1-7). Hershey, PA: IGI Global.

Long, S.D., Picherit-Duthler, G., & Duthler, K.W. (2009). Managing relationships in virtual team socialization. In M. Khosrow-Pour (Ed.), Encyclopedia of information science and technology (2nd ed., pp. 1-7). Hershey, PA: IGI Global.

Lurey, J.S., & Raisinghani, M. S. (2001). An empirical study of best practices in virtual teams. *Information & Management, 38*, 523-544.

Majchrzak, A., Rice, R., Malhotra, A., King, N., & Ba, S. (2000). Technology adaptation: the case of a computer-supported inter-organizational virtual team. *MIS Quarterly, 24*, 569-600.

Markus, M.L. (1994). Finding a happy medium: Explaining the negative effects of electronic communication on social life at work. *ACM Transactions of Information Systems, 12*, 119-149.

Maznevski, M.L., & Chudoba, K.M. (2000). Bridging space over time: Global virtual team dynamics and effectiveness. *Organization Science, 11*, 473-492.

McLeod, P.L., & Liker, J.K. (1992). Electronic meeting systems: Evidence from a low structure environment. *Information Systems Research, 3*, 195-223.

Metiu, A. (2006). Owning the code: Status closure in distributed groups. *Organization Science, 17*, 418-435.

Mullen, S. (2005). Media choice, interpersonal relationships, and problem solving. *Proceedings of the IADIS Virtual Multi Conference on Computer Science and Information Systems*, 444-451.

Murray, S.R., & Peyrefitte, J. (2007). Knowledge type and communication media choice in the knowledge transfer process. *Journal of Managerial Issues, 19*, 111-133.

Olson, G.M., & Olson, J.S. (2000). Distance matters. *Human-Computer Interaction, 15*, 139-179.

Ouelette, J.A., & Wood, W. (1998). Habit and intention in everyday life: The multiple processes by which past behavior predicts future behavior. *Psychological Bulletin, 124,* 54-74.

Pauleen, D. J. (2003). An inductively derived model of leader-initiated relationship building with virtual team members. *Journal of Management Information Systems, 20,* 227-256.

Pauleen, D.J. (2003-2004). An inductively derived model of leader-initiated relationship building with virtual team members. *Journal of Management Information Systems, 20,* 227-256.

Peiro, J.M., & Melia, J.L. (2003). Formal and informal interpersonal power in organisations: Testing a bifactorial model of power in role-sets. *Applied Psychology: An International Review, 52,* 14-35.

Pena, J., Walther, J.B., & Hancock, J.T. (2007). Effects of geographic distribution on dominance perceptions in computer-mediated groups. *Communication Research, 34,* 313-331.

Perrow, C. (1986). *Complex organizations: A critical essay* (3rd ed.). New York: McGraw-Hill.

Pfeffer, J. (1992). *Managing with power.* Boston: Harvard Business School Press.

Powell, A., Piccoli, G., & Ives, B. (2004). Virtual teams: A review of current literature and directions for future research. *The DATA BASE for Advances in Information Systems, 35,* 6-36.

Rains, S.A. (2005). Leveling the organizational playing field—virtually: A meta-analysis of experimental research assessing the impact of group support system use on member influence behaviors. *Communication Research, 32,* 193-234.

Rains, S.A. (2007). The impact of anonymity on perceptions of source credibility and influence in computer-mediated group communication. *Communication Research, 34,* 100-125.

Raven, B. H. (1992). A power/interaction model of interpersonal influence: French and Raven thirty years later. *Journal of Social Behavior and Personality, 7,* 217–244.

Reicher, S.D., Spears, R., & Postmes, T.T. (1995). A social identity model of deindividuation phenomena. *European Review of Social Psychology, 6,* 161-198.

Robert, L.P., & Dennis, A.R. (2005). Paradox of richness: A cognitive model of media choice. *IEEE Transactions on Professional Communication, 48,* 10-21.

Rogelberg, S. R., Luong, A., Sederberg, M. E. & Cristol, D. S. (2000). Employee attitude surveys: Examining the attitudes of noncompliant employees, *Journal of Applied Psychology, 85,* 284–293.

Rosen, B., Furst, S., & Blackburn, R. (2007). Overcoming barriers to knowledge sharing in virtual teams. *Organizational Dynamics, 36,* 259-273.

Saunders, C.S., Robey, D., & Vaverek, K.A. (1994). The persistence of status differentials in computer conferencing. *Human Communication Research, 20,* 443-472.

Siegel, J., Dubrovsky, V., Kiesler, S., & McGuire, T.W. (1986). Group process in computer-mediated communication. *Organizational Behavior and Human Decision Processes, 37,* 157-187.

Silver, S.D., Cohen, B.P., & Crutchfield, J.H. (1994). Status differentiation and information exchange in FtF and computer-mediated idea generation. *Social Psychology Quarterly, 57,* 108-123.

Sosik, J.J., Avolio, B.J., & Kahai, S.S. (1997). Effects of leadership style and anonymity on group potency and effectiveness in a group decision support system environment. *Journal of Applied Psychology, 82,* 89-103.

Sosik, J.J., Avolio, B.J., Kahai, S.S., & Jung, D.I. (1998). Computer-supported work group potency and effectiveness: The role of transformational leadership, anonymity, and task interdependence. *Computers in Human Behavior, 14,* 491-511.

Spears, R., & Lea, M. (1994). Panacea or panopticon? The hidden power in computer-mediated communication. *Communication Research, 21,* 427-459.

Sproull, L., & Kiesler, S. (1986). Reducing social context cues: Electronic mail in organizational communication. *Management Science, 32,* 1492-1512.

Suler, J. (2004). The online disinhibition effect. *CyberPsychology & Behavior, 7,* 321-326.

Tajfel, H., Billig, M.G., Bundy, R.P., & Flament, C. (1971). Social categorization and intergroup behavior. *European Journal of Social Psychology, 1,* 149-178.

Tajfel, H., & Turner, J. C. (1986). The social identity theory of intergroup behaviour. In S. Worchel & W. G. Austin (Eds.), *Psychology of intergroup relations* (2nd ed., pp. 7-24). Chicago: Nelson-Hall.

Tan, K.W.P., Swee, D., Lim, C., Detenber, B.H., & Alsagoff, L. (2008). The impact of language variety and expertise on perceptions of online political discussions. *Journal of Computer-Mediated Communication, 13,* 76-99.

Tidwell, L.C., & Walther, J.B. (2002). Computer-mediated communication effects on disclosure, impressions, and interpersonal evaluations: Getting to know one another a bit at a time. *Human Communication Research, 28,* 317-348.

Vecchio, R.P. (1997). Power, politics, and influence. In R.P. Vecchio (Ed.), *Leadership: Understanding the dynamics of power and influence in organizations* (pp. 71-99). Notre Dame, IN: University of Notre Dame Press.

Wagner, D.G., & Berger, J. (1993). Status characteristics theory: The growth of a program. In J. Berger & M. Zelditch, Jr. (Eds.), *Theoretical research programs: Studies in the growth of theory* (pp. 23-63). Stanford, CA: Stanford University Press.

Wagner, D.G., & Berger, J. (2002). Expectation states theory: An evolving research program. In J. Berger & M. Zelditch, Jr. (Eds.), *New directions in contemporary sociological theory* (pp. 41-76). Lanham, MD: Rowman & Littlefield Publishers.

Walker, H.A., Thye, S.R., Simpson, B., Lovaglia, M.J., Willer, D., & Markovsky, B. (2000). Network exchange theory: Recent developments and new directions. *Social Psychology Quarterly, 63,* 324-337.

Walther, J.B. (1992). Interpersonal effects in computer-mediated communication: A relational perspective. *Communication Research, 19,* 52-90.

Walther, J.B. (1995). Relational aspects of computer-mediated communication: Experimental observations over time. *Organization Science, 6,* 186-203.

Walther, J.B. (1996). Computer-mediated communication: Impersonal, interpersonal, and hyperpersonal interaction. *Communication Research, 23,* 3-43.

Walther, J.B., & Burgoon, J.K. (1992). Relational communication in computer mediated interaction. *Human Communication Research, 19*(1), 50-88.

Walther, J.B., Slovacek, C.L., & Tidwell, L.C. (2001). Is a picture worth a thousand words? Photographic images in long-term and short-term computer-mediated communication. *Communication Research, 28,* 105-134.

Warkentin, M., & Beranek, P.M. (1999). Training to improve virtual team communication. *Information Systems Journal, 9,* 271-289.

Weber, M. (1946). *From Max Weber: Essays in Sociology* (H.H. Gerth & C. W. Mills, Eds.). New York: Oxford University Press.

Weisband, S., Schneider, S.K., & Connolly, T. (1995). Computer-mediated communication and social information: Status salience and status differences. *Academy of Management Journal, 38,* 1124-1151.

Willer, D., Lovaglia, M.J., & Markovsky, B. (1997). Power and influence: A theoretical bridge. *Social Forces, 76,* 571-603.

Witmer, D.F., Colman, R.W., & Katzman, S.L. (1999). From paper-and-pencil to screen-and-keyboard: Toward a methodology for survey research on the Internet. In S. Jones (Ed.), *Doing Internet Research* (pp. 145-162). Newbury Park, CA: Sage.

Yoo, Y., & Alavi, M. (2004). Emergent leadership in virtual teams: what do emergent leaders do? *Information and Organization, 14,* 27-58.

Yukl, G., & Falbe, C.M. (1990). Influence tactics and objectives in upward, downward, and lateral influence attempts. *Journal of Applied Psychology, 75,* 132-140.

Yukl, G., & Tracey, J.B. (1992). Consequences of influence tactics used with subordinates, peers, and the boss. *Journal of Applied Psychology, 77,* 525-535.

Yukl, G., Chavez, C., & Seifert, C.F. (2005). Assessing the construct validity and utility of two new influence tactics. *Journal of Organizational Behavior, 26,* 705-725.

Yukl, G., Guinan, P.J., & Sottolano, D. (1995). Influence tactics used for different objectives with subordinates, peers, and superiors. *Group & Organization Management, 20,* 272-296.

Zhang, S., & Fjermestad, J. (2006). Bridging the gap between traditional leadership theories and virtual team leadership. *International Journal of Technology Policy & Management, 6,* 274-291.

Zigurs, I. (2003). Leadership in virtual teams: Oxymoron or opportunity? *Organizational Dynamics, 31,* 339-351.

Zigurs, I., Poole, M.S., & DeSanctis, G.L. (1988). A study of influence in computer-mediated group decision making. *MIS Quarterly, 12,* 625-644.

TABLE 1

Availability of Status Characteristics in Various Virtual Team Communication Channels

Communication Channels[5,6,7]	Status Characteristics			
	Race	Gender	Age	Abilities, Skills, or Expertise
FtF	X	X	X	X
Telephone & Audio Conferencing		X	X	X
Video Conferencing	X	X	X	X
Chat Rooms (for text interactions)		X		X
File Transfer		X		X
Virtual Reality (e.g., virtual reality meeting rooms in Second Life)[8]	X	X	X	X
Handoff Collaboration (e.g., using the Tracking Changes option in MS Word)		X		X
Simultaneous Collaboration (e.g., Windows Meeting Space, Google Chrome, or Timbuktu)		X		X

---

[5] Communication channels listed have been adapted from Olson & Olson (2000).

[6] Some channels of virtual team communication allow for FtF as well as computer-mediated communication (e.g., meeting room video and audio conferencing) while other channels are not intended for this purpose (e.g., desktop video and audio conferencing). The table above reflects the status characteristics and status cues allowed for by the computer-mediated portion of the communication.

[7] Text-only channels (e.g., chat rooms, file transfer, handoff collaboration) will only convey the communicators' genders if they include their names.

[8] Avatars may inaccurately portray the communicator's race, gender, and age in virtual reality settings. Nonetheless, team members may form expectations based on these status characteristics.

TABLE 2

Availability of Status cues in Various Virtual Team Communication Channels

Communication Channels	Status Cues							
	Posture	Intonation	Style of Dress	Gestures	Fluency of Speech	Frequency of Speech	Eye Contact	Typing Speed
FtF	X	X	X	X	X	X	X	
Telephone & Audio Conferencing		X			X	X		
Video Conferencing	X	X	X	X	X	X	X	
Chat Rooms (for text interactions)								X
File Transfer								
Virtual Reality (e.g., virtual reality meeting rooms in Second Life)	X	X	X	X	X	X	X	
Handoff Collaboration (e.g., using the Tracking Changes option in MS Word)								
Simultaneous Collaboration (e.g., Windows Meeting Space, Google Chrome, or Timbuktu)					X	X		X

TABLE 3

Demographic Data

Participant ID	Female	≤ 40 Years Old	≤ 18 Month Tenure in Virtual Team	Junior Team Member	Low Technological Expertise
Helen	X	X	X	X	
Richard		X	X		
Charles		X	X	X	
Donna	X	X	X	X	
Kimberly	X				
William		X	X	X	
Michelle	X		X	X	
Linda	X			X	X
Laura	X	X	X		X
Daniel				X	X
Dorothy	X	X	X	X	
Robert					
Betty	X	X	X	X	
Patricia	X				X
Karen	X				
Elizabeth	X	X		X	
Carol	X			X	
Sandra	X		X		
Sharon	X		X	X	
Michael		X		X	
Susan	X	X		X	
Joseph			X		X
Christopher		X		X	

Note. All participant IDs are pseudonyms.

TABLE 4

Traditional Influence Tactics also Found in Virtual Teams[9]

Code	Definition
Rational Persuasion	You use logical arguments and factual evidence to persuade the person that a proposal or request is practical and likely to result in the attainment of task objectives.
Pressure	You use demands, threats, frequent checking, or persistent reminders to influence the person to do what you want. Pressure includes the sub-codes of following up, frequent communication, forwarding a previous communication as a reminder, guilt, and brute force.
Legitimating Tactics	You seek to establish the legitimacy of a request by claiming the authority or right to make it, or by verifying that it is consistent with organizational policies, rules, practices, or traditions.
Inspirational Appeals	You make a request or proposal that arouses enthusiasm by appealing to the person's values, ideals, and aspirations, or by increasing the person's confidence that he or she would be able to carry out the request successfully.
Ingratiation	You seek to get the person in a good mood or to think favorably of you before making a request of proposal (e.g., compliment the person, act very friendly).
Exchange	You offer an exchange of favors, indicate willingness to reciprocate a favor at a later time, or promise the person a share of the benefits if he or she helps you accomplish a task.
Consultation	You seek the person's participation in planning a strategy, activity, or change for which you desire his or her support and assistance, or you are willing to modify a request or proposal to deal with the person's concerns and suggestions.
Coalition Tactics	You seek the aid of others to persuade the target person to do something, or use the support of others as a reason for the target person to agree to your request.
Personal Appeals	You appeal to the person's feelings of loyalty and friendship toward you when you ask him or her to do something.

---

[9] Definitions of Traditional Influence Tactics are taken from Yukl et al (1995)

TABLE 5

Technologies Used to Adapt Traditional Influence Tactics

Code	Definition
Sharing Your Screen	You share your screen with the target so that she or he has a better understanding of what you are asking of her or him
Using Electronic Whiteboard Technology	You use the whiteboard feature of LiveMeeting in order to collaborate with others.
Sending Urgent Zen Mail	You send email messages with the entire message in the subject line. Emails with the subject beginning 'Urgent' are very influential.
Using Emoticons	You use emoticons in order to soften the influence request. Emoticons convey friendliness and a more personal tone. Related to the following codes: Personal Tone (communication style) & Personal Appeals (FtF influence tactic).
Highlighting Important Information	You highlight important information in email communications and other document to ensure that it gets read.
Polling	You enable the polling feature of emails or LiveMeeting in order to involve team members' in the planning of a project.

TABLE 6

Ambiguity Reduction Techniques

Code	Definition
Information Sharing	You provide the target with details needed to either accomplish the task or that convey the importance of or reason for the request.
Creating Accountability	You hold the target accountable for the request, typically in a public setting (e.g., on a conference call).
Giving an Example	You provide the target with an example of what you want her or him to do for you in order to ensure the desired outcome. This action may or may not stem from the fact that the target does not know how to perform the task.

TABLE 7

Upward Influence Tactics

Influence Tactic	Successful	Unsuccessful
Traditional Influence Tactics		
Coalition Tactics	0	1
Inspirational Appeals	1	0
Legitimating Tactics	4	2
Pressure	10	2
Rational Persuasion	6	0
Total	21	5

TABLE 8

Lateral Influence Tactics

Influence Tactic	Successful	Unsuccessful
Traditional Influence Tactics		
Consultation	1	2
Exchange	5	0
Inspirational Appeals	1	0
Legitimating Tactics	4	7
Pressure	6	7
Rational Persuasion	1	1
Ambiguity Reduction Technique		
Create Accountability	1	1
Give Example	3	2
Total	22	20

TABLE 9

Downward Influence Tactics

Influence Tactic	Successful	Unsuccessful
Traditional Influence Tactics		
Consultation	1	0
Exchange	1	0
Ingratiation	1	0
Inspirational Appeals	2	0
Legitimating Tactics	0	2
Pressure	6	3
Rational Persuasion	3	0
Ambiguity Reduction Technique		
Give Example	1	1
Information Sharing	1	0
Total	16	6

TABLE 10

Strategic Use of Technology to Get Attention

Code	Definition
Using Multiple Communications	You use several different communication technologies to ensure that your target gets your request. Typically the additional technologies are used for follow-ups, but multiple communications can also be part of the initial request.
Using the Target's Technology of Choice	You communicate with your target using her or his preferred communication technology so that she or he will be more receptive to your request.
Using IM to Check Availability	You ensure that your target is able to receive your influence request immediately.

TABLE 11

The Influence Strategy of Building Relationships

Code	Definition
Establishing Trust	Developing confidence in someone. Being able to rely or depend on someone.
Getting to Know Others	Learning information about others.
Putting a Face with a Name	Being able to relate nonphysical attributes of a person with her or his physical appearance in order to better understand the person.
Team Building	Strengthening the relationship of all team members.
Building a Sense of Community	Creating a collective feeling of belonging to a group.

TABLE 12

Status Cues

Code	Definition
Body Language	Nonverbal communication such as gestures, facial expressions, and posture.
Amount of Talk	The amount a person speaks or asserts her or himself during a conversation.
Taking Initiative	Being the person who begins a line of communication.
Maturity of Voice	The age a person sounds when she or he speaks.
Interrupting	Speaking while someone else is speaking. Talking over someone else.
Volume	The loudness of a person's speech.

TABLE 13

Diffuse Status Characteristics

Code	Definition
Age	Status differences associated with the biological age of a person. In virtual teams, age is associated with comfort with technology, a distinction between old-school and new-school, and organizational tenure.
Occupational Position	Status differences associated with a person's position based on either their profession (e.g., physician) or their organizational position (e.g., Vice President of Marketing).
Gender	Status differences associated with the gender of a person.
Race & Ethnicity	Status differences associated with a person's biological race or ethnicity.
Parenthood	Status differences associated with whether or not a person has a child.
Education	Status differences associated with the amount of formal education a person has had.

TABLE 14

Specific Status Characteristics

Code	Definition
Knowledge	Status differences associated with what a person knows that is relevant to the job. Knowledge reflects a broader understanding than expertise.
Expertise	Status differences associated with what a person knows that is relevant to the job. Expertise reflects a deeper understanding than knowledge.
Tech Savviness	Status differences associated with a person's ability to use technology to its fullest extent.
Communication Skills	Status differences associated with how a person communicates that are relevant to the job.
Job Skills	Status differences associated with what a person can do that is relevant to the job.

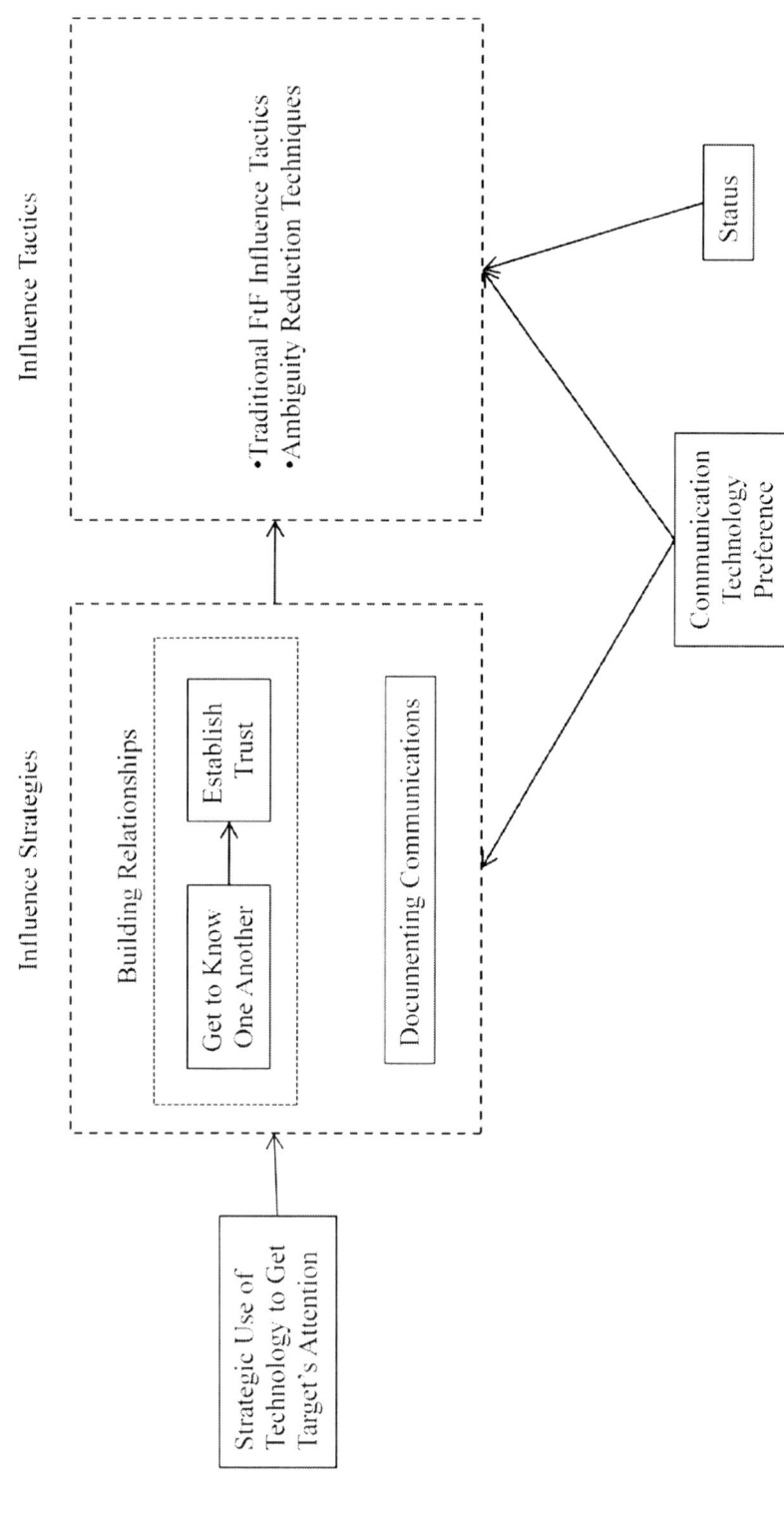

FIGURE 1. *Influence Process in Virtual Teams*

APPENDIX A: PRE-INTERVIEW QUESTIONNAIRE

1. How long have you been a member of your current (or most recent) virtual team?

   (in months) _____

2. What is your gender?

   ○ Male

   ○ Female

3. What is your ethnicity?

   ○ Asian American

   ○ Black/African American

   ○ Latino/Hispanic

   ○ White/Caucasian

   ○ Other: [_____]

4. What is your age? (in years) _____

5. How many people are on your virtual team? (numeric value) _____

6. How many members of your virtual team are in your building? (numeric value)

   _____

7. How many time zones does your team span? (numeric value) _____

8. How many times <u>per year</u> do all of the members of your virtual team meet FtF?

   (numeric value) _____

9. Do you have a team leader?

   ○ Yes

○ No

10. If so, was the leader formally assigned by the organization?

○ Yes

○ No

APPENDIX B: INTERVIEW PROTOCOL

1. Team-related Items (Interview Warm Up/Context Questions)

    a. Tell me a little about your virtual team.

        i. Describe the structure of your team (e.g., hierarchy vs. flat).

    b. How did you become part of the virtual team (e.g., legacy, hybrid)?

    c. Describe your position in the virtual team.

    d. On what type of project is your virtual team working?

    e. What technologies do you and your team members use to…

        i. …communicate with one another?

        ii. …share documents?

        iii. …collaborate?

    f. How often (percentage wise) do you use these technologies?

        i. More than others, etc.?

2. Influence-related Items

    a. Sometimes people have more work to do than they have time to get it done. Other times a task arises that they simply do not want to do. Or perhaps information is needed that is hard to come by. For these reasons, and many others, people often seek to use their influence to get by what want. I am interested in those situations which lead people who work in virtual teams to influence members of their virtual teams.

    b. Describe an incident in which you succeeded in getting a virtual team member to do something you wanted that pertained to your job. For example, did you have a deadline approaching and need help from a team member to complete the project?

    i. Tell me about your co-worker.

        1. How similar are they to you?

        2. How different are they from you?

        3. How did you decide you were going to ask him/her as opposed to another team member?

    ii. How do you normally communicate with one another?

    iii. What did you want from your co-worker?

    iv. Did you have a plan or strategy on how you would get what you wanted?

        1. Did it involve using the technology available to your virtual team?

    v. How did you use technology to get what you wanted? I am going to ask you to give me examples of specific things you did (e.g., emoticons, delaying messages, scheduling meetings, tracking changes in a document, tracking availability through instant messaging, etc.)?

    vi. What happened next? (How did your co-worker respond to your attempt?)

        1. How did he/she use technology to respond?

    vii. In thinking about the end result, how close was it to the outcome you hoped for?

1. If it wasn't right on target for what you were looking for, why do you think it wasn't? (e.g., coworker wasn't as invested, situational constraints, etc.)
   viii. Would you have done the same thing if you had been interacting with the person face-to-face?
c. Describe an incident in which you succeeded in getting a [either equal or lower status – depending on what they described in question 'a'] virtual team member to do something you wanted that pertained to your job. For example, did you have a deadline approaching and need help from a team member to complete the project?
   i. Tell me about your co-worker.
      1. How similar are they to you?
      2. How different are they from you?
      3. How did you decide you were going to ask him/her as opposed to another team member?
   ii. How do you normally communicate with one another?
   iii. What did you want from your co-worker?
   iv. Did you have a plan or strategy on how you would get what you wanted?
      1. Did it involve using the technology available to your virtual team?
   v. How did you use technology to get what you wanted? I am going to ask you to give me examples of specific things you did (e.g.,

emoticons, delaying messages, scheduling meetings, tracking changes in a document, tracking availability through instant messaging, etc.)?
- vi. What happened next? (How did your co-worker respond to your attempt?)
    1. How did he/she use technology to respond?
- vii. In thinking about the end result, how close was it to the outcome you hoped for?
    1. If it wasn't right on target for what you were looking for, why do you think it wasn't? (e.g., coworker wasn't as invested, situational constraints, etc.)
- viii. Would you have done the same thing if you had been interacting with the person face-to-face?

d. Describe an incident in which you failed in getting a virtual team member to do something you wanted that pertained to your job. For example, did you have a deadline approaching and need help from a team member to complete the project?
- i. Tell me about your co-worker. (elicit status characteristics of co-worker)
    1. How similar are they to you?
    2. How different are they from you?
    3. How did you decide you were going to ask him/her as opposed to another team member?
- ii. How do you normally communicate with one another?

iii. What did you want from your co-worker?

iv. Did you have a plan or strategy on how you would get what you wanted?

   1. Did it involve using the technology available to your virtual team?

v. How did you use technology to get what you wanted? I am going to ask you to give me examples of specific things you did (e.g., emoticons, delaying messages, scheduling meetings, tracking changes in a document, tracking availability through instant messaging, etc.)?

vi. What happened next? (How did your co-worker respond to your attempt?)

   1. How did he/she use technology to respond?

vii. Why do you think you were unsuccessful in your attempt?

viii. Would you have done the same thing if you had been interacting with the person face-to-face?

e. Describe the person in your virtual team who has the most influence. This person is not necessarily the person formally in charge of your team. For example, oftentimes unexpected tasks pop up during a project. The person who decides who will work on these tasks may be the most influential team member. Thinking about the most influential person on your virtual team...

   i. Why are they so influential?

   ii. How do they get other people in the team to do what they want?

   iii. How do they use technology to get what they want?

CPSIA information can be obtained at www.ICGtesting.com
Printed in the USA
BVOW07s1410010414

349413BV00010B/358/P